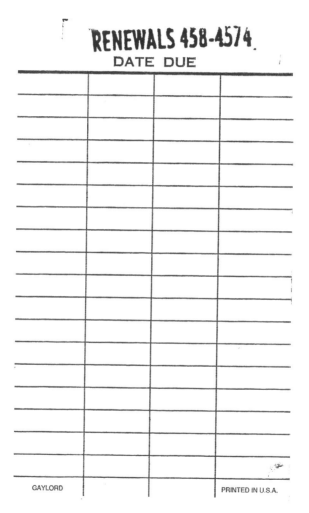

RENEWALS 458-4574

DATE DUE			
GAYLORD			PRINTED IN U.S.A.

Japan and the Internet Revolution

Japan and the Internet Revolution

Ken Coates
College of Arts and Science
University of Saskatchewan
Canada

and

Carin Holroyd
College of Commerce
University of Saskatchewan
Canada

First published 2003 by
PALGRAVE MACMILLAN
Houndmills, Basingstoke, Hampshire RG21 6XS and
175 Fifth Avenue, New York, N.Y. 10010
Companies and representatives throughout the world.

PALGRAVE MACMILLAN is the global academic imprint of the Palgrave Macmillan division of St Martin's Press, LLC and of Palgrave Macmillan Ltd. Macmillan® is a registered trademark in the United States, United Kingdom and other countries. Palgrave is a registered trademark in the European Union and other countries.

ISBN 0–333–92153–4

This book is printed on paper suitable for recycling and made from fully managed and sustained forest sources.

A catalogue record for this book is available from the British Library.

Library of Congress Cataloging-in-Publication Data

Coates, Kenneth, 1956–
    Japan and the Internet revolution/Ken Coates and Carin Holroyd.
        p. cm.
    Includes bibliographical references and index.
    ISBN 0–333–92153–4 (cloth)
        1. Information technology–Economic aspects–Japan. 2. Internet–Economic aspects–Japan. 3. Internet–Japan. I. Holroyd, Carin. II. Title.
HC465.I55C63 2003
384.3′0952–dc21

                                                                        2003048065

10 9 8 7 6 5 4 3 2 1
12 11 10 09 08 07 06 05 04 03

Printed and bound in Great Britain by
Antony Rowe Ltd, Chippenham and Eastbourne

To the memory of Klaus Pringheim, in appreciation of his gracious encouragement of our work and his infectious love for Japan.

To the memory of Roger Boisvert, a Canadian entrepreneur who played a crucial role in the introduction of the Internet to Japan.

We miss them both.

Contents

List of Tables

Preface

Japan and the Internet Revolution seeks to explain Japan's unique and fascinating entry into the Internet age. After lagging years behind North America, Europe and other industrial regions in identifying the opportunities presented by the Internet, the country rallied around the new technology in the mid-1990s and forged a most unusual 'networked' society. Private use of the Internet is now more commonplace in Japan than, by some standards, any place on earth. Tens of millions of Japanese surf the web via their telephones. A crammed commuter car in Osaka or Tokyo typically hosts a dozen or more users of the mobile Internet, or *keitai*. While personal computer user remains comparatively small, Japan consumers are capitalizing on broadband installations, navigating through traffic in Internet-connected vehicles, working with machines connected to the Internet, or shopping in vibrant e-commerce and m-commerce markets.

The idea for this book arose during our first extended visit to Akihabara (Electric Town) in Tokyo. This small corner of one of the world's most fascinating cities had surrendered itself to the digital age. As is described in the Introduction, Akihabara is a digital 'temple', marketing computers and related electronic products with a verve and ferocity not evident in most other countries. One of us is a self-declared Internet junkie, devoted to staying at the forefront of the digital revolution and intensely fascinated by the marriage of the Internet and national culture which has occurred in Japan. The other, a casual user of the Internet but a specialist in Japanese business and political economy, was intrigued by the evident gap between that encountered in Akihabara and portrayals in the western media of a national economy caught in the grips of despair and deflation and lacking the innovative capacity to join the 'New Economy'.

Thus began this exploration of the digital revolution in Japan and, more specifically, of the evolution of the Internet in the Land of the Rising Sun. The digital phenomenon is so recent and so complex that this book aims to provide an overview of a diverse and rapidly changing Internet environment, striving to offer context and a general analysis rather than the kind of detailed statistical, technical, economic and sociological investigations which will inevitably be done. The book seeks to draw the broad outlines of a transformation in process and to thereby highlight the critical changes and developments currently underway in Japan.

   Detailed analytical work on the Internet in Japan is at the very early stages. Japan's digital transformations have not been well-covered in newspapers, magazines, e–zines (digital magazines), and the like. Bill Gates and other North American promoters of the Internet world are known the world-over; Masayoshi Son is not. The efforts by Amazon.com and Dell to create a new e-commerce economy are lauded internationally for creativity and innovation. Rakuten's Japan-focus Internet market is, in contrast, virtually unknown outside the country. The world celebrates the creative accomplishments of Microsoft, Nokia, Sun Microsystems, Cisco and other companies which have played critical roles in the promotion of personal computer and Internet use; general understandings of the remarkable success of DoCoMo's wireless Internet are minimal at best. The economic opportunities associated with the Internet, particularly before the dot.com melt-down in 2001–2002, generated professional analysis of technological developments, market demand and Internet usage. These studies have declined over the past while, in keeping with the diminished commercial enthusiasm for dot.com stocks and new technology companies. The so-called 'grey literature' of government reports and NGO commentaries are diverse and extensive, including numerous contributions by various government agencies, but they report more what is planned than what has actually been accomplished. Academic analysts, particularly those from outside Japan, have been slow to turn their attention to the study of the Internet and its impact on Japanese society. The existing work has been important in our analysis, as will be evident in the following chapters, and have used the general audience media to gain insights into the diverse and complex role that the Internet is playing within Japan. The study of the Internet, in Japan and elsewhere, suffers somewhat from the hype (and even hysteria) that accompanied the dot.com boom and the rhetoric about the New Economy. This revolution has been at once oversold and underestimated. Much of the early writing had a boosterish flavour, celebrating relatively minor achievements and foreshadowing major and permanent transformations. It is intriguing, even nostalgic, to read these 'visionary' assessments of only a few years back, and to assess the degree to which promoters and critics alike overestimated the immediate impact of the Internet while, simultaneously, paying too little attention to some wide-ranging and more subtle transitions. It is a challenge to avoid the traps and pitfalls that are so easy to stumble into during the early phases of a major transformation. There are profound and important transformations underway in Japan (and elsewhere); equally, the manner in which the country responded to the digital revolution reveals a great deal about the nation, its people, and the nature of technological innovation and change.

Technological adaptation, adoption, and implementation reflect national, regional, and ethnic cultural values and considerations. One sees, for example, the inherent conservatism of the national government in the slow and plodding response to the development of the Internet overseas. There are glimpses of Japanese isolation and paranoia in the early concerns about the intrusion of the Internet into Japanese cultural life. The reaction of the keitai (mobile Internet) and other innovations speaks directly to the suitability of specific technologies to the unique socio-economic situation in Japan – and provides strong evidence of the technological and commercial creativity of Japanese business. There is evidence, too, of Japan's strong desire to reproduce its stunning post-1970s global dominance in key industrial sectors, particularly in the international roll-out of the mobile Internet. In these and other ways, Japan's Internet revolution is distinctive and is reflective of national conditions and key elements of Japanese culture. There is, of course, the other side of the Internet – the instantaneous, globalizing influences of the digital revolution. If Japan is a networked nation, it exists within a networked world, where the rapid flow of data and information holds the potential to alter the assumptions and realities of the industrial age. Japan has changed a great deal from the isolated, inward-looking nation of the early 19th century; how it adjusts to the dramatic, transformative capabilities of the digital age remains, in large part, to be seen.

It is a challenge, perhaps even misguided folly, to attempt a general analysis of the cultural, economic, political and social 'revolution' that is in its formative stages. Japan's adaptation of the Internet raises many questions and provides, at this early date, relatively few answers. For people outside Japan, the contours of the country's digital transformation remain a mystery. Japanese technology firms – Sony, Masushita, NEC, Fujistu, Canon and many others – are known globally for the technological refinements and consumer-focused innovations. But despite these digital accomplishments, the perception remains that Japan continues to lag behind in responding to the potential of the Internet. *Japan and the Internet Revolution* has a simple goal: to alert readers outside the country to the complex, innovative, incomplete, uneven, creative, and technologically refined implementation of the

Internet that is currently underway in Japan. In charting a broad and sometimes sweeping course, moving away from detailed technical analysis to consider very general questions about the relationship between culture and technology, this book should provide readers with a non-technical, accessible overview to contemporary developments in Japan's digital transformation.

Ken Coates, College of Arts and Science
Carin Holroyd, College of Commerce
University of Saskatchewan
Saskatoon, Saskatchewan
Canada

Acknowledgements

It is always a great delight to acknowledge the contributions of friends and colleagues who played a role in the completion of this book. We were encouraged to undertake this project by Ken Henshall, a fellow Japan specialist and friend from New Zealand. Much of the research and writing was completed when Carin was teaching at Kansai-Gaidai University in Japan. The generous support and assistance provided by the Centre for International Education, and particularly Dean Yamamoto and Reiko Hiromoto, is particularly appreciated. We benefited enormously from the advice and suggestions of Dr. Paul Scott. We tried out many of the ideas in this book on our good and patient colleagues in the Japan Studies Association of Canada; we greatly appreciate their supportive feedback.

Work undertaken earlier, under the direction of Klaus Pringsheim and Martin Thornell of the Canada-Japan Trade Council, helped spark an interest in e-commerce and the development of the Internet. Klaus was a giant of a man, both in physical size and in terms of his impact on the study of Japan in Canada. We were aided enormously by the kindness and insights of Roger Boisvert, founder of Global Online in Tokyo, who spent long hours explaining the intricacies of the early days of the Internet in Japan and whose passion for the Internet and Japan kindled our interest in this project. His tragic death at the hands of a robber in Los Angeles saddened friends and colleagues around the world. Various professional associates contributed a great to the development of this book. Sarah Deverane of the Office of the Dean of Arts, University of New Brunswick at Saint John, provided invaluable assistance with the early stages of the research. The staff in the office of the Dean, College of Arts and Science, and the Provost and Vice-President (Academic), University of Saskatchewan were particularly gracious and incredibly patient in dealing with the many drafts of this project. We want to draw particular attention to the assistance provided by Sharon Ford and Jennifer Simpson. Amber McCuaig, Ken's associate in the Provost's Office, went far beyond the call of duty in helping to pull this book together. We must thank, as well, the good folks at Palgrave-Macmillan, whose patient support of this project has been greatly appreciated, and our editor, Ann Marangos. Grateful thanks also go to the Social Sciences and Humanities Research Council of Canada for Carin's generous post-doctoral research fellowship.

Our families remain, as always, a source of inspiration and encouragement, particularly during the challenging times that occurred during the writing of this book. Two children joined our family over the last two years. Hana Beth came to us in Hanoi, Vietnam in August 2001 and has been a dazzling ray of happiness since her arrival. Substantial portions of this book were written in Curitiba, Brazil, our temporary home while we finalized the adoption of our very happy and enthusiastic son, Marlon Leslie. Thank-you to them and to Beth, Les and Lisa Holroyd, Marge Coates, Colin Coates and Megan Davies, Janice and Terry Mori, Bradley, Mark and Laura Coates.

KEN COATES and CARIN HOLROYD
Saskatoon
Saskatchewan

Introduction: The Tip of the Digital Iceberg

It is hard to separate substance from glitter in the fast-paced world of dot.coms, Internet 'revolutions', New Economies, e-government, telehealth, web usage rates, mobile telephony and the thousands of other developments, issues, technologies and processes associated with the digital transformation. Beginning in the early 1990s, when the World Wide Web emerged from highly specialized technical circles into global prominence, the Internet-based revolution has seen both remarkable highs and shocking lows. New ideas have gained widespread currencies and the core assumptions which seemed so safe only a few years ago have been proven to be faulty. Far beyond the false promise of massively inflated stock prices, however, lays a transformation of fundamental importance. Around the world, and in Japan no less than other leading industrial nations, thousands of transitions are underway as governments, businesses, organizations, political parties, individuals and societies seek to come to terms with the potential, the opportunities and the intrusions of the Internet.

Japan is a high-energy, fast-moving, dynamic nation, characterized by the frantic dash to the trains and subways by salarymen and office ladies and frenetic workers who greet the dawn on their way to the office and who last well into the evening. The country's stores glitter with advertising, bright lights and endless promotions. Urban landscapes are dominated by blazing corporate logos and the frantic flickering of the pachinko parlors. Every second person seems to be speaking into a *keitai* (mobile phone), even on crowded subways and in busy downtown restaurants. High-tech toys and animation share children's affections with the gentler, Hello Kitty, attractions of the previous decade. The old traditions and pre-industrial arts remain in place – for

Japan's true accomplishment has been to marry the ancient and the modern – but the country seems locked into a high-tech frenzy.

Western commentators have, for almost a decade, intoned worried thoughts about the future of Japan's economy.[1] In the aftermath of the fabled bubble economy, when sky-rocketing real estate values made a square mile in downtown Tokyo worth more than all of the privately held property in Canada and when Japanese companies ventured off shore with a vengeance to buy up foreign holdings, there seemed to be almost palpable relief when the recession hit. Books that proclaimed and explained Japan's rise to economic superpowerdom surrendered shelf space to volumes documenting the collapse of the Japanese 'miracle' and forecasting the painful reckoning that lies ahead. Rising national debt, spectacular corporate failures (sometimes marked by the suicide of major business leaders), a slow retreat from selected international markets, and continued confusion in national politics all signaled, it seemed, the dawn of a less prosperous economic era in Japan. The 'new' Japan, forecasters suggested, would be older, less dynamic, weaker and with a declining standard of living. While the country would not retreat into poverty, the story went, Japan seemed destined to lose its place as a world leader.

Reconciling these two images of Japan is difficult. Japan is not simply an Asian version of western or American popular culture. The bright lights and high energy are not superficial glitter, without deep and profound connections to the national economy and political system. And if Japan has fallen into the kind of deep economic funk described by international analysts, it is not apparent at the street level, where consumerism runs unrestrained and the work ethic that drove the country's post-World War II revitalization remains everywhere in evidence. The macro-economic statistics do not lie, but they may mislead. Japan's government is running a massive annual budget deficit. But the country also has huge trade surpluses with the United States and most of Asia, maintains a strikingly high national savings rate, has enormous capital reserves (and is the world's largest creditor nation), and has (despite the years of economic difficulty) maintained one of the world's highest per capita incomes and highest standards of living.

For people who grew up in the post-World War II era, when the phrase 'Made in Japan' was synonymous with commercial junk, the idea of Japan as a nation of high-technology innovators still rests uncomfortably. The next generation, raised with Sony transistor radios, Honda motorcycles and Toyota cars, saw Japan as a great imitator, taking North

American and European products, making them more affordable, reliable and consumer-friendly. The past decade presents a different image, of companies like Toshiba, Softbank and NEC, tackling the largest computer and electronics firms in the world and often coming out on top. 'Made in Japan' now means high quality, cutting edge innovation and futuristic styling. It is the country of anime and robotic dogs, portable computers and the Sony Walkman. As the world struggles to understand Japan, there is also a generational disconnect to the public questioning and global impressions of this most puzzling of countries.

Japan and the Internet Revolution seeks to come to terms with the impact on Japan of high-technology and the digital 'revolution.' It examines the Japanese response at all levels, from consumers to business, from government to industry, and from the workplace to the world of politics. This work does not align itself with either extreme camp in the global effort to explain the country's current economic and social system. Japan is in serious economic difficulty, we believe, but beyond the harsh news from the banks, insurance companies and government departments, there are signs of enormous vitality and creativity. Similarly, analysts who suggest that the country is prepared for continued global leadership are a tad optimistic, for it remains to be seen if the societal and political structures which underlay Japan's remarkable post-war recovery are well-suited for the unique demands of the digital age, which place a priority on global corporate activity and technological synergies. What is clear, however, is that Japan is re-inventing itself once again. The Meiji period saw Japan leap, with remarkable verve and tenacity, from the pre-industrial world to Asian economic pre-eminence in a single generation. After World War II, indulged and supported by its former American enemy, Japan emerged from the ruins of conflict to become again a major industrial power. Through the 1960s and 1970s, Japan redesigned itself as a leading technological nation, and developed business models and innovation systems soon copied the world over. After the Plaza Accord financial restructuring of the 1980s, the country shook off the effects of the currency revaluation and retooled its economy for Asian leadership and cutting-edge production techniques.

Now, Japan faces yet another challenge, its most profound in fifty years. The tools, approaches and structures of the 'old' economy, all of which served Japan exceptionally well, seem ill-suited for the speed and prerogatives of the 'new' economy. In an age of dot-coms, collapsed product cycles, e-commerce, and the information revolution, Japan finds itself saddled with top-heavy corporate systems, a slow-moving bureaucracy that seems out of touch with contemporary realities, an ossified political structure focused primarily on the necessities of re-election, and a conservative social order that favours stability over change. The country has resources, to be sure, including one of the world's best education systems, an enormous capacity for technological innovation, the globe's most curious consumers, and the capacity to experiment, based on substantial and continued national and personal wealth. Japan sits, too, at the eastern edge of the formidable Asian market, favoured by location and culture but burdened by its history in its attempts to provide leadership and direction to the millions of consumers in the emerging economies of East Asia. Anticipating the future, never an easy task, is made all the more difficult in the Japanese situation, where uncertainty about the present, deeply entrenched assumptions about Japan's capacity to change, and the complexities of the global and Asian markets all affect the country's prospects. *Japan and the Internet Revolution* argues, however, that the country is coping with its internal and external challenges better than most observers understand and is capitalizing on the opportunities of the digital age in ways that suggest a promising, if not exciting, future for the land of the digital rising sun.

Understanding Japan's response to the information and computer revolution requires a visit to Akihabara. This commercial district in the northern end of Tokyo, the post-World War II black market for radio parts, is one of the commercial and technological marvels of the modern age. One goes to the Silicon Valley in California to get a feel for the hyper-energetic, risk-taking, speculative fever which sparked the internet boom in North America, to Bangalore, India, to get a flavour for the emergence of a developing world response to western industrial domination, and to Ireland to gauge the capacity of the 'new' economy to rewrite historical rules and assumptions about economic marginality. In same vein, anyone seeking to explore the depth of the consumer-driven fascination with usable technology heads to Akihabara. Stepping off the train at one of the two main stations in the area, visitors immediately see signs pointing to 'Electric Town' (written, incidentally, in English as well as Japanese). During regular business hours, the trains empty out at these stops, one of the most popular shopping venues in the country. Young people, in particular, pour out of the station and onto the streets of one of the most impressive electronic shopping Meccas in the world.

The Akihabara district is about twenty blocks, give or take a few, in size. It stretches along three main streets, offers over 500 stores, spills over into the surrounding side-streets and back allies. The electronic immersion experience begins immediately upon exiting the station. Walls of the surrounding high-rises are plastered with flashy advertisements for new products. The ubiquitous Japanese tissue people stand in abundance on the streets outside. (Many Japanese companies provide free packages of tissue paper as advertisements, and one rarely travels through a large station or major shopping district without being offered several packages of tissue paper.) The square immediately in front of the station is typically occupied by small dealers, hawking tourist type items and giving demonstrations – like in a North American carnival – of household products. Announcements and exhortations pour out of street-side speakers, occasionally even offered in Mandarin as a inducement to foreign visitors. The traffic flow attracts advertisers of all sorts – over 200,000 shoppers on the average weekend and some 80 million a year – and the surrounding buildings are festooned with brilliantly lit corporate slogans and posters. Companies staging special promotions in Akihabara usually send scantily dressed young girls and corporate mascots to the station. While the girls screams out enticements over loud-speakers, the cuddly mascots race toward small children, and others hand out brochures and product information. (*Gajin*, or foreigners, are often ignored by those offering enticements; these efforts are aimed almost exclusively at domestic consumers.)

Large corporate stores dominate the Akihabara district. LAOL, one of the biggest electronic dealers in the country has several large operations in the area, each of them six or seven stories high. The stores themselves are marvels of technological over-kill. Each floor is dedicated to a specific product – from air conditioners and home computers to digital cameras and DVD players – or to specific companies, including Sony, Toshiba, Fujistu, Apple and IBM. In a tip of the hat to foreign visitors – and Akihabara is one of the easiest places in all Japan to meet foreigners – the major stores devote a floor or large section to duty free merchandise or goods specially designed for foreign markets. The foreign goods sections are rarely as interesting or attractive as those aimed at Japanese consumers, but they often offer products not yet available in most international markets.

But Akihabara is more than a concentration of large electronics stores. In fact, it is a microcosm of Japanese commerce, and offers insights into the country's economic complexity. Alongside the major stores, competing on price, service or, most often it seems, the loudness of their store front enticements, are a whole range of smaller shops. Some of these are several stories in size, although much smaller than the big companies and rather like a rabbit warren inside. Others specialize in a single product or product line, offering DVDs, computer games, mobile telephones, household appliances or specialized goods. Companies launching new products often rent a sizable store front in Akihabara; Gateway (a South Dakota-based customized computer company) has used such a presence to introduce itself to Japanese consumers. Dell Computers has two outlets in the area, selling between 20 and 30 computers a day in 2000. At the opposite end of the scale, in long, narrow passageways running through the large buildings, are dozens of tiny stalls, again offering electronic parts, repair services, and smaller versions of electronics stores. Whereas the large stores offer flashily dressed young men and women, hawking the latest technology, these stalls are typically staffed by a single older man, offering a limited product line with none of the commercial panache seen in the bigger stores.

Akihabara sprawls out into the surrounding area, and filters up off street level. In less prime locations, companies offering demonstrations or specialized services have set up shop. Further away, along the back alleys, shoppers find unique deals. Often business people who have secured a special order or an end of product run of an consumer item will take over an empty store and sell their goods from tables or even the back of a pick-up truck. *Caveat emptor* applies in Japan as anywhere! Two major Akihabara dealers, Two Top and Rocket Inc., ended 2000 in deep financial difficulty; Rocket Inc's revenues had fallen by over one-third in ten years. Second-hand computer stores, repair shops, and store-fronts offering one-time deals fill up the remaining areas. Throughout the district, smaller operations set up in second, third or fourth floor offices. Some of these companies, like Two Top, are national chains, specializing in buying discounted goods from major manufacturers and offering them at cut-rate prices; Two Top's profit margin was estimated to be between 2–3 per cent. In these stores, as in the alley ways and private-run shops, bargaining is part of the commercial culture. Stock changes as soon as products run out, ensuring a steady turn-over of both goods and customers. Prices vary enormously from the high-end stores to the back of the pick-up pedlar, and shoppers scurry from shop to shop in search of low prices.

The atmosphere is frenetic as companies vie for a portion of the almost $40 billion in annual sales. Loudspeakers blare endlessly. Neon lights sparkle and blink without a break. Rock music can be heard from a dozen stores at any given time. Street-side hucksters proclaim the benefits of their store and product line, in polite Japanese fashion filled with *sumimasen* and *domo arigato* (excuse me and thank you). Mascots and promotional girls cajole and entice. Crowds build around any new product demonstration. Walls of television monitors, each attached to a computer games controller, attract hundreds of eager young Japanese shoppers, anxious to see the latest offerings from the software producers. Akihabara was the obvious choice for the launch of the Sony Playstation II in March 2000, with the company's 2 million units selling out the first weekend. The following week, crowds returned en masse again, lining up patiently in front of the same stores, this time to have a chance to purchase the new games just released for the Playstation II. Rumours and stories spread rapidly through the lines, sending shoppers (particularly those near the end) rushing from store to store in search of better opportunities.

With support from the Tokyo Metropolitan Government, Akihabara is being prepared for an expanded role. Worried about the potential loss of customers to on-line sales, growing competition from off-site locations (led by companies like Kojima, Bic Camera and Yodobashi Camera), and the potential impact on consumers of the shift from PC-based to mobile computing, the Akihabara district is being touted as a major high tech development zone. The Metropolitan Government is making available a large (over 8 acre) site in the district – one of the largest development sites in the area – for a research/convention complex, designed to solidify the area's place in Japan's IT revolution. Other elements in the aggressive initiative include a training centre, warehouse and product development units. Promoters tout the combination of high tech competence and the ready accessibly to Japanese consumers, a potentially critical package for software and hardware developers. According to the Tokyo government, the attractions of the area include 'strong name recognition in the global IT industry, an established magnet for young, tech-savy shoppers,; and convenient location with easy access to railroads and airports'.[2]

This is not the Japan of legend, of tea-houses and geishas, samurai and temples. And this is not the Japan of contemporary western mythology, of over-worked salarymen, bored housewives and hopelessly stressed students. This is the essence of contemporary Japan, young, aggressive, male and female, wired, motivated, capitalistic and creative. This the showpiece of some of the world's most creative consumer electronics companies and the stomping grounds of the most open minded and experimental consumers. Akhibarara is where DoCoMo promotes I-mode phones and where rivals seek new customers for their versions of the internet-capable mobile units. This is not the place where visitors see last year's American products being introduced to the Japanese, but rather where afficionados of electronic innovation come to see new items one or two years before they appear in North American or European stores.[3] (Sony Playstation II was launched, to considerable acclaim and a few glitches, in Japan in the spring of 2000 and in North America in the fall.) Akihibara reveals the complexity and depth of the consumer-drive information revolution in a way that may be unmatched in the world, for one encounters here everything from the latest portable phones to electronically monitored toilets, from mobile telephones to Internet-capable microwaves.

Osaka, Japan's second-largest city, has a similar commercial zone, called Den Den (Electric Electric) Town. It pales in comparison to Akihabara and consists of several crowded streets, the sidewalks not able to handle the press of consumers, offering a similar variety of products and services. Like its Tokyo counterpart, Den-Den Town includes major department stores, smaller shops, and tiny stalls, but the effect is dramatically different. Unlike Akihabara, which is as much an entertainment centre as a shopping district, Den-Den Town is small, crowded and strictly for sales. There is little opportunity to circulate, comparison shop and sample the diversity of Japanese electronic devices. The whole area is about one fifth the size of Akihabara, just as the Kansai's 6.5 million people represent about 20 per cent that of the greater Tokyo area. Den-Den Town is more a large congregation of shops than a consumer test zone, although its crowds and high-energy provide abundant evidence that Japanese interest in consumer electronics is not limited to Tokyo.

Akihabara leaves a stunning impression. It is diverse, capitalistic to its core, aggressive, experimental, consumer-oriented, and extremely energetic. That the stores cover the range from major department stores and commercial outlets for most of the world's biggest names in consumer electronics to single owner shops and tiny repair stalls reveals the complexity of Japan's commercial structure – often ignored but perhaps one of the country's key sources of strength. The barrage of advertising, overwhelming on first visit, illustrates the aggressiveness and competitiveness that drives Japanese innovation; whatever competition Japanese firms feel from overseas is but a fraction of what they typically experience from companies within the country. The hands on, in your face, advertising efforts make North American promotional efforts seem tame by comparison and render European tactics positively genteel.

What is most compelling is the fact that it is not all glitter and noise. Behind the competitive façades and promotional campaigns lies a wealth of industrial power, product innovation, and global manufacturing ability. And underlying these commercial attributes rests a firm foundation of national loyalty and determination, a strong (if still flawed) education system, and a work ethic that sustains the country's standard of living and competitive position. In Akihabara, one gets to sees the tip of the Japanese digital iceberg, the spot where the behemoth that has been helping propel the global technological revolution emerges into public view. While there is smoke, mirrors and loud music, there are also products galore, many of which will never appear outside the Japanese market and some of which will, like the Sony Walkman, eventually become standard fare around the world. Akihabara affords an excellent introduction to the realities of Japan in the age of the digital revolution.

Exploring Japan's adaptation to the Internet revolution requires the national equivalent of an extended trip to Akihabara. It is crucial to look at what is on the surface, and to assess such things as consumer, business and government computer use, the state of the Japanese-language internet, and the extent of electronic commerce and technological innovation. But the search must go much deeper and consider the nation's preparedness for the information age and whether or not Japan has the political will to adapt to new economic and technological realities. It is crucial to step back from the contemporary fray and examine the manner in which Japan transformed itself from a nation of heavy industry and cheap labour to the world's leading producer of robots and high-technology products, for in this history there are clues to the challenges and opportunities facing the country. Such an investigation must also consider, in a comparative, international dimension, the intersection of culture and technology. Is Japanese society and the country's value system well-suited to the electronic revolution, or will the attributes that once brought prosperity and world leadership hold the country back in the future? (For our purposes, we define 'culture' as the series of learned behaviours, values and structures which define a particular people. This learning focuses on the development of language skills, which help define the boundaries between groups and influence relationships within the group. Culture also includes the web of social and economic relationships which influence personal interactions, family composition and function, institutional development at the social and political level, the evolution of economic systems and institutions and the relationships that individuals hold to these structures. Culture also refers to the habits, customs, rituals, duties, responsibilities and other aspects which, evolving over time, have become integral to the self-definition and annual and life cycle activities of the members of a particular society.) The question of the intersection of cultural characteristics and techno-economic developments, ultimately, are critical issues which will shape all efforts to understand the emergence and prospects of Japan in the information age. As one journalist observed, Japan has entered a 'Maybe Restoration', a period of potential transformation that will determine the country's position in the age of Internet competition.[4]

Technology, the Japanese miracle and the internet: scholars' evaluations:

Japan's experience with digital technology and the Internet has not gone unnoticed among analysts and academics.[5] While the western business press has, generally, been critical of the country's overall performance and surprisingly silent on the extent and nature of the revolution in mobile Internet, scholars have long been fascinated by Japan's adaptation and adoption of industrial and commercial technologies. In fact, as many researchers have noted, the rapid acceptance of new approaches to manufacturing and business processes fueled the Japanese 'miracle' and assisted the country's rise to international economic prominence. Scholars have, on an international scale, been fascinated by the relationship between technology and economic development, and continued to struggle with the implications – made more immediate and potentially disruptive by the launching of the Internet-based economic transformation – of rapid technological change.[6]

In the early days of the technology boom, when the emphasis rested on computer hardware and related products and not software and dot.com innovations, many analysts predicted Japan's ultimate victory in the global struggle for IT dominance. Tom Forester, an Australian academic, wrote ominously (and incorrectly) in 1993:

> The entire high-tech sector – including growth areas such as biotechnology, new materials, pharmaceuticals and medical equipment – is at risk from Japanese domination unless America and Europe make dramatic changes to their industry, trade, and technology policies, as well as their industrial structrure and indeed their whole approach to business and wealth creation. Unless the West

learns the lessons of Japan's high tech business strategy and changes course, there is a grave danger that America and Europe could become little more than industrial museums – and Japan's economic triumph will be complete.[7]

Books and articles appeared by the dozens, lauding Japan's commercial success and foreshadowing an era of Japanese domination of the global economy.[8] Japanese corporations and business–government relations were held out as models for international emulation, and celebratory accounts of the unbroken success of Japanese firms appeared regularly.[9]

Analysts following Japan's uneven and rocky economic road in the 1990s and early years of the 21st century have struggled to understand the continuing role of technological innovation. Alfred Chandler's study of the evolution of the consumer electronics industry, *Inventing the Electronic Century*, gives pride of place to a discussion of the critical role played by Japanese companies, particularly Sony, in creating consumer interest in this sector.[10] Eamonn Fingleton, a Tokyo-based journalist specializing in industrial and financial matters, argued in his less than prescient *Blindside: Why Japan Is Still on Track to Overtake the U.S. by the Year 2000* that Japan's emphasis on technological excellence is the fundamental underpinning of national success.[11] His more recent and more predictive study, *In Praise of Hard Industries: Why Manufacturing, Not Information Technology, Is the Key to Future Prosperity*,[12] criticizes the west's headlong rush into the dot.com environment and urges governments to maintain an emphasis on manufacturing prowess. Drawing on his earlier work, Fingleton emphasizes Japan's culture of technological and process innovation in suggesting that the country's continued commitment to manufacturing excellence will stand the national economy in good stead in the future. He argued, at the height of the Internet euphoria, that Japan was right not to make a leap of faith in the world of information technology, advice which turned out to be quite appropriate.

Adam Posen suggests, in a study of the link between technological innovation and economic performance, that too much has been made of the uniqueness of Japanese conditions in the 1960–mid-1980s period. He argues that fairly normal preconditions for economic success, including high savings rates, sizeable investment in research and development, strong entrepreneurship and a high level of competition fuelled Japan's economic rise. He attributed Japan's economic slide in the latter years of the 20th century to a weak culture of academic and industry innovation and research, limited venture capital and the absence of support for start-up companies. He noted widespread concern in government and the private sector about the country's comparatively poor performance in leading edge science, particularly in biotechnology, and information technology, and the impact of out-dated regulations and commercial laws on corporate innovation and competition.[13] He joins with a multitude of western voices in arguing that the key to Japan's future economic success lies more in adjusting political, administrative and financial regulations to globally competi-tive norms than in restarting or overhauling the country's approach to innovation.

There was ample reason for economic uncertainty in the early 1990s, and it was not until the advent of the 'new economy' and the Internet-fueled frenzy in the latter half of the decade that Japan's continued emphasis on manufacturing prowess over commercial innovation appeared wrong-headed. By the dawn of the new century, less than a decade after analysts trumpeted the 'inevitability' of Japan's economic success and applauded the accomplishments of Japan Inc (government–industry cooperation in economic development), scholars were decrying Japan's inability to respond to new realities. In *Can Japan Compete?*, one of the most important and influential critiques of Japanese economic behaviour, Michael Porter, Hirotaka Takeuchi and Mariko Sakakibara offered a critical analysis of the country's economic prospects:

> What we have found is that almost none of the conventional wisdom is true. Japan's much celebrated bureaucratic capitalism is *not* the cause of Japan's success; in fact, it is most closely associated with the nation's failures ... [T]he core of the problem is that the government mistrusts competition and therefore is prone to inter-vene in the economy in ways that harm the nation's productivity and prosperity. The received wisdom about Japan's past corporate success has more merit, but it is dangerously incomplete. What was once a viable approach to competition no long works in today's global marketplace. Hampered by their own approach to competing, Japanese companies undermine their own profitability.[14]

Critical appraisals of Japan have now become commonplace, as schol-ars endeavour to understand the collapse of the Japanese bubble and the country's inability to return to economic pre-eminence.[15] Just as the laudatory books over-reached their mark, so too do most of the critical assessments find too much at fault in a country which, despite its economic difficulties, remains at the forefront in innovation and standard of living.

Analysts have struggled, and not just on matters relating to the Internet and Information Technology, to analyse the changing nature of Japanese innovation and technological advance. Until recently, a great deal of emphasis was placed, appropriately, on the work of the Ministry of International Trade and Industry and, largely as a result of Fingleton's work, the Ministry of Finance, in fostering technological innovation. The close association of business and government referred to by Porter and his colleagues was long cited as the cornerstone of Japan's economic advantage and was, for a time, presented as a model for national economic development. Detailed studies of the information technology sector,[16] including *Computers Inc.: Japan's Challenge to IBM*, by Marie Anchordoguy and Koji Kobayashi's *The Rise of NEC*,[17] illustrate the fundamental importance of this partnership in the early days of digital development.[18] In the most important study in this field, *The Market and Beyond: Cooperation and Competition in Information Technology Development in the Japanese System*, Martin Fransman documented the emergence of a 'system' of technological innovation which shifted over time and which involved substantial internal cooperation. This investigation pre-dated the advent of the commercial Internet and emphasized manufacturing and research processes. Fransman illustrated the manner in which telecommunications companies, in particularly, worked with university and government researchers to explore opportunities for technological advancement.[19]

Japanese companies continue to attract considerable attention from international analysts, as has been the case for much of the last twenty years. The development of 'Just In Time' and Total Quality Management manufacturing processes, particularly by Toyota and Nissan, have drawn numerous scholarly investigations. Ken-ichi Imai's investigation of the process and structural roots of Japanese innovation is a good example of the strong interest in this field.[20] A series of important studies by Michael Cusumano have documented the application of technology-based solutions to manufacturing challenges and described how these innovations effected corporate operations.[21] Cusumano's most important study, an investigation of software manufacturing in Japan, examines the manner in which major corporations (he focuses on Hitachi, Toshiba, NEC and Fujitsu) managed the development of major computer programs, the foundation of technology-based manufacturing, retailing and ecommerce. Through his study, he demonstrates the formidable challenges facing individual companies and consortia and argues that, while problems and challenges remain, Japanese companies have been able 'to make this technology less dependent on a small number of highly skilled engineers and to achieve efficiencies across a series of projects rather than building software from scratch for each customer or job order'.[22] His analysis of the application of factory processes and design systems to the development of software is presented as something of a warning to overseas business people, whom he argued have prematurely dismissed the Japanese as a commercial threat in the software sector.

The rapid expansion of the Internet and the growth of the digital economy encouraged scholars to turn their attentions to the evolution of networked societies. Manuel Castells, one of the leading commentators on turn of the century global transformations, wrote of the current era of rapid change:

> The core of the transformation we are experiencing in the current revolution refers to technologies of information processing and communication. Information technology is to this revolution what new sources of energy were to the successive Industrial Revolutions, from the steam engine to electricity, to fossil fuels, and even to nuclear power. ... What characterizes the current technological revolution is not the centrality of knowledge and information, but the application of such knowledge and information to knowledge generation and information processing/communication devices, in a cumulative feedback loop between innovation and the uses of innovation.[23]

Castells' study of the 'network society' is one of many attempts to understand the significance of the IT revolution for the world. He uses Japanese examples, particularly from the manufacturing sector, to describe how networks are becoming increasingly pervasive and economically and socially influential. Castells endeavours to address regional and national-specific cultural responses to the Internet and digital technology, arguing that values, internal structures, government-business dynamics, attitudes toward change and other factors play critical roles in determining the short and long-term impact of the information revolution. Like many others, however, he asserts that the primary implications of the new technologies are global in scope, reach and impact, and foreshadow an era where nations matter less and technology plays an increasingly vital role in mediating relationships.

A great deal of the analytical effort attached to the information revolution has focused on economic and commercial aspects of the

Internet. An enormous effort, revealed by the collapse of the dot.com boom, was based on shaky assumptions and over-enthusiasm. The Internet, and particularly retail adaptations to e-commerce, simply could not support the hopes and dreams of the high technology innovators.[24] Much the same has been true in other sectors, including e-government, media convergence (the union of television, radio, newspapers and other media distribution systems), tele-health and technology-based education.[25] In none of these sectors have the visions of the promoters or the hostile reactions of critics[26] been borne out. In most countries, including Japan, analysts have described the early shortcomings of the Internet-based revolution and raised doubts about the long-term impact of the digital information systems.

The following chapters describe and assess the digital transformations of the past two decades, focusing primarily on the more recent expansion of Internet-based activity in Japan. Chapter 1 offers a brief overview of the development of the computer industry in Japan and considers the manner in which the country positioned itself in the first wave of the digital revolution. Considerable attention is devoted to both the culture of innovation that surrounded corporate and government initiatives and to the limited adoption of computer-based activities within the country. Chapter 2 examines the approach of the national government to the opportunities and challenges presented by the Internet. It describes the early resistance to the Internet, the cumbersome regulatory procedures and laws which slowed progress in the area and the initiatives led by computer scientist Jun Murai to overcome a raft of administrative hurdles. This chapter also examines the post-2000 conversion of the governments of Prime Ministers Yoshiro Mori and Junichiro Koizumi, the former a reluctant advocate and the latter a keen enthusiast, to the belief that expansion in Internet use was central to the country's economic prospects.

The middle chapters in this book focus on the economic and commercial aspects of the digital revolution in Japan. Chapter 3 looks a peculiarly Japanese phenomenon, the *keitai* revolution, involving the lightening fast development of the mobile Internet in the country. The startling emergence of DoCoMo and the establishment of a viable m-commerce economy represents the world's most successful commercial use of the Internet. The manner in which the *keitai* reflects key characteristics of Japanese society and has become imbedded in Japanese culture within a short period of time illustrates the adaptive capacity of the Internet, the creative abilities of Japanese industry and business, and the powerful influence that national characteristics exert over technological innovation. Chapter 4 focuses on the evolution of e-commerce in Japan, with particular emphasis on the role of the Softbank empire of Masayoshi Son (Japan's version of Bill Gates). As a late entrant into the game, and constrained by a conservative financial regime, Japan largely missed out on the early euphoria of the dot.com boom in North America. Because it did not join in the initial excitement and because the Japanese approach to venture capitalism is more restrained than that in the United States and elsewhere, Japan did not suffer from the serious dislocations associated with the bursting of the dot.com bubble. At the same time, there have been impressive and unique implementations of e-commerce in Japan, at both the business to business and business to consumer level. Japan's culture of creative capitalism, with its sharp focus on customer service, has found some ways to capitalize on the Internet, and has shown a particular gift for using the web to meet uniquely Japanese challenges and opportunities.

The final two chapters provide a broader context within which to understand and situate Japan's engagement with the Internet. Chapter 5 explores a series of key elements of the country's use of the Internet. In addition to the presentation of a variety of statistical analyses of aspects of Internet and computer use, this chapter examines the manner in which Japan presents itself to the world via the Internet. It explores, in particular, the crucial issue of language use and the Internet, in which Japan has managed to develop the largest and most successful nation-specific Internet presence in the world. Chapter 6 places Japan's Internet activity and transformation within a global context. It addresses, in a preliminary way, what we believe to be a critical long-term question about the relationship between national cultures and the opportunities presented by the Internet. We provide an assessment of the unique manner in which Japan has implemented and exploited the Internet and a comparative assessment of how other countries have confronted the digital revolution. Based on these evaluations, we argue that a country like Japan may prove to be extremely well situated to capitalize on the economic, political, social and cultural aspects of the Internet. As will become clear in Chapter 1, Japan's Internet involvement did not emerge in a vacuum. Instead, the country's active engagement in the development of computers and consumer electronics laid a foundation which proved of critical importance in determining the subsequent shape of the digital revolution in Japan.

1
Uneasy Steps: Japan and the Development of the Digital Society

Fifty years ago, only the most visionary dreamer would have conceived of Japan as being at the cutting edge of the digital revolution. In the early 1980s, it would have been a relatively safe bet to guess that Japan would be a world leader in the evolution of computer-based technologies and consumer electronics. Both assumptions, ironically, would prove to be wrong. Through the 1960s, Japan emerged as a major innovator in retail and industrial technologies and played a critical global role in introducing the world to the opportunities posed by transistors. Having established itself as the leading producer of consumer electronics and as major player in the emerging field of personal computers, the country nonetheless reacted very slowly to the potential of the Internet and delayed the nation-wide implementation of this crucial technology. Continuing the unusual pattern of leading and then falling behind on technological innovation, Japan again reverted to its position in front of other nations in terms of the mobile Internet, becoming the first nation to establish cell phone-based Internet services as the key element in national innovation and transformation.

Japan still suffers from a lingering 1960s stereotype. As the country rebuilt from the ravages of World War II, it re-emerged on the world's economic stage as an industrial imitator. Japanese companies produced a small number of original products; instead, they perfected processes of reverse engineering. Firms identified high demand foreign products, developed Japanese models, and reintroduced them to world markets at a lower price. This strategy, which drew on the nation's formidable work ethic, low wages, and collective determination, underlay the surge in economic growth through the 1960s.

Short-term returns, however, were offset by the creation of a peculiar image of Japan and Japanese innovation. The country, the foreign assessment went, did not invent new products. They were not innovative on the design and development side, only on the production and management elements of manufacturing. Japan, observers claimed, cherry-picked discoveries and inventions from other countries, bought up patents and licensing rights, and found ways to turn expensive products into accessible consumer items. It was a critical role, to be sure, but it focused on low levels of research and design, high investments in process engineering, and sharp attention to foreign consumer markets. It worked. Japan's industrial and economic growth from the late 1950s through to the early 1970s was enviable, by any international standard. Surprisingly, however, industrialists, inventors and business observers the world over were not particularly impressed.

Observers, typically pointing to the comparatively small number of patents and Nobel Prize winners in Japan, attributed the lack of creativity to the Japanese education system. Japan's schools and universities, the argument went, produced excellent technicians (particularly engineers and industrial scientists), but did not create a culture of innovation. The intensity and rigidity of the examination-based high school system forced students to memorize, underlay a phenomenal national work ethic, and provided a superb scientific and literacy base for the population as a whole. But, the assertion went, Japanese education did not create free thought, creativity, and the quirky, irreverent curiosity that pushes an innovation culture. (The United States was typically held up as the contrary example, despite the nation-wide mediocrity of science education and with insufficient attention to the fact that many 'American' inventors were foreign-educated immigrants, attracted by the financial opportunities and risk-oriented investment climate of America.) There was, as with most stereotypes, an element of truth to the image of Japan's educational inflexibility, but the steadfast assertion that the country lacked a culture of innovation and creativity was overstated.

The legacy of this industrial development lingers on, particularly in the computer field, although the country is getting more and more kudos for its consumer-oriented innovations and its product development.[1] (The hosting of a major commercial exhibit of Japanese products and corporate innovations, held in London, England in the autumn of 2000, by Bartle Bogle Gearty advertising agency, is one indication of the growing recognition.) With many commentators assuming and asserting that the United States maintains a hammerlock on industrial and consumer invention and development, casual observers of Japanese manufacturing continue to assume that the country's industrialists are imitative, slow-moving, and process-obsessed. The country receives little credit for the development of new products, for new scientific and technical discoveries, and for its efforts to create ideas for the future. This image did not emerge from nowhere, however. A review of the origins of Japan's computer revolution helps explain both the origins and persistence of the stereotype. It also illustrates the extraordinary steps taken by the Japanese government and Japanese industry to overcome its slow start in the computer industry and the country's considerable achievements since the 1960s.

The roots of Japan's contemporary technological accomplishments are deeply imbedded in the country's history. In the years before the opening of Japan to the outside world in the mid-19th century, Japan developed a sophisticated, highly cultured society, with a strong record for technological innovation. By cutting itself off from most of the world, however, Japan skipped the early stages of the industrial revolution and, when western traders arrived en masse in the late 19th century, had little beyond silk and a few resource products to offer in exchange. Aware of the striking technological gap, the nation launched into the Meiji Restoration, an astonishingly comprehensive and rapid industrialization of the country. Within a few short decades, Japan adopted railways, industrial factory production, and science-based education. The government moved aggressively to bring Japan to the forefront of the industrial age and achieved its goals by the early years of the 20th century. Much of this success came from emulating western accomplishments in manufacturing and production, dispatching young scholars to North America and Europe to secure the necessary education and training, and bringing in foreigners to teach new technologies and explain scientific discoveries. Japan offered little that was new to the industrial mix of this era, but proved remarkably adept at copying western science and technology and adapting it to Japanese circumstances.

Contrary to long-standing assumptions, however, Japan did not simply rest on its ability and willingness to seek inspiration outside its borders. A greatly expanded public education system upgraded rapidly the academic attainments of the population at large. Universities modelled, in part, on western institutions emerged and expanded quickly, soon rivalling some of the best universities in the world. The large *zaibatsu* (business conglomerates that were the precursors of the contemporary *keiretsu*) invested in product development and industrial processes. Japan's economic and technological advances, showcased in its bold and aggressive moves toward Korea, China and Southeast Asia in the 1930s and the blistering attack on American, Dutch and British military bases in 1941, owed a great deal to a national culture of innovation and a desire for industrial and economy autonomy. While the rest of the world knew very little about Japanese developments – language remaining an even more formidable barrier than cultural attributes until well after World War II – the lack of awareness only underscored the west's inability to comprehend the nature and extent of Japan's advances.

The ravages of World War II and the slow emergence from the Allied occupation left a country economically crushed and emotionally drained. Having wrestled with the demons of militarism, and determined to rebuild their battered nation, the Japanese turned the same energies and resolve that had made them a military power to the reconstruction of their economy. Until the early 1960s, the country's advantage lay largely in an impressive national work ethic, the government's willingness to clear the way for commercial development, and the workers' willingness to place their needs and interests behind those of the company and the state. The result was an extraordinary turn around of the Japanese economy, and the steady rebuilding of national confidence. By the mid-1960s, in a land which had seemingly sacrificed quality of life and the health of the environment on the mantle of economic progress, Japan had re-established itself as a major economic power. The best and most dramatic was yet to come.

While the arrival of high-quality Japanese automobiles and motorcycles on international markets in the late 1960s and 1970s signalled publicly the 'arrival' of Japanese manufacturing, a more important transition had started before this time. Japanese companies, particularly Sony, were among the first to recognize the potential of newly developed transistors. Before these American innovations came along, televisions and radios operated with vacuum tubes. These tubes made the new consumer electronics bulky and expensive; overheating and power surges often resulted in blown tubes. Japanese firms recognized, before their North American competitors, that transistors could be adapted to replace the vacuum tubes, thus reducing the size and cost of electronic entertainment devices and improving their efficiency and reliability. The Japanese had not invented the transistors, but they had recognized and harnessed the potential of the new technology.[2]

In the process of exploiting the commercial potential of the transistor, the Japanese re-invented themselves economically. To this point, Japan exported cheap products (toys, inexpensive clothing, and the like), counting on inexpensive labour to make their goods competitive.

With transistors, Japanese companies shifted their focus, providing innovative consumer items that were of high quality, inexpensive compared to their competitor's products, and fashionable. In the early 1960s, Japanese manufacturers were widely dismissed as mediocre and unreliable; within a decade, the products by the same companies were viewed with admiration and respect. Japan had done more than secure the right to use an important technology – transistors; in the process, the country's industries discovered the key to national economic success: a commitment to quality, product innovation, responsiveness to consumers, and a collaborative management system based on co-operation between workers and management. The rest of the industrialized world began to pay more attention to Japan, watching the once impoverished country overcome the formidable liability of World War II guilt and defeat to emerge as a major economic power. Competitors noted, in particular, the innovative management system, the strong customer orientation of manufacturers, and the growing emphasis on technological innovation.

But in one key sector – computers and computerization – Japan remained well behind the United States, the world's leading nation in the field. (The following description of the development of the Japanese computer industry and the role played by government in getting it off the ground was drawn primarily from Marie Anchordoguy's fascinating book, *Computers Inc: Japan's Challenge to IBM*.)[3] Japan was not alone in under-estimating the importance of computers; leading American firms and key officials were, as late as the 1970s, dismissing the possibility of widespread commercial and personal use of computers. As a consequence, a few powerhouses, led by IBM, had the field much to themselves. And, on the foundation of IBM's dominance, the United States had a formidable advantage in what had the potential to be a crucial industrial, commercial and consumer market. Faced with corporate reluctance to enter the manufacture of computers and computer components (no Japanese company had the money for such a venture), and realizing that each passing year added to IBM's and America's considerable advantage, the Japanese government launched a concerted national program designed to force-feed a computer industry and to give the country a foothold in this promising sector.

In 1960, a group of public and private sector representatives decided that a domestic computer industry was in Japan's strategic long-term interests – no one wanted to be dependent on foreign companies for computers – and that the private sector would need help from the

Ministry of International Trade and Indusrtry (MITI) to be successful. MITI agreed and it believed that the first step in developing this new Japanese industry was to offer an environment in which companies would be willing to invest and consumers willing to buy. The Japanese government did this, first, by putting in place various protectionist measures (tarrifs were raised to 25 per cent) designed to limit the number of machines foreign companies could sell in Japan and to restrict foreign investment. Secondly, they offered subsidies, low inter-est loans and loan guarantees to give companies access to the capital they would need to begin computer production. Thirdly, the govern-ment implemented policies whose objective was to avoid 'some of the wasteful aspects of competition – cutthroat price wars, redundant R&D, and uneconomic scales of production – while still requiring that they compete in the final marketplace. These policies influenced the market structure of the industry and the behavior of the computer firms in two primary ways: They reduced the costs and risks of entering and operating in the computer business to encourage firms to make the heavy investment necessary to become competitive; and they reduced the number of companies operating in each market segment to help firms gain economies of scale in R&D and production. Hammered out by the public and private sector, these policies were generally struc-tured in ways that did not completely shelter the firms from competi-tion and that required them to make better products to survive in the long term'.[4]

MITI negotiated the acquisition of IBM's patents on behalf of Japanese industry thus ensuring that the Japanese companies did not compete against each other to obtain them, thereby bidding up the price. MITI then gave the patents to seven Japanese companies – Fujitsu, Hitachi, NEC, Mitsubishi, Toshiba, Oki and Matsushita – which all began producing computers commercially. (Matsushita withdrew a few years later so only six firms remained.) While MITI protected its domestic makers, the ministry encouraged vigorous competition among them and promoted the sector on a national and international scale. In addition, the threat that the government would soon open up the computer market to foreign makers (which the United States was pressuring Japan to do) made the Japanese companies keenly aware that protectionism was only buying them time and that they would have to become internationally competitive as quickly as possible.

Through various means ranging from import quotas to coercion, MITI ensured that a domestic market existed for the Japanese produc-ers. As an import license from MITI was required to buy a foreign computer, MITI was able to strongly encourage Japanese companies who wished to buy a foreign computer to buy a Japanese one even though the domestic product, particularly in the early 1960s, was nowhere near as good. When social pressure to buy Japanese was not enough, 'there were even cases where we had to make it compulsory for them to change (their minds) from a foreign to a domestic computer', recalled a former MITI bureaucrat.[5] Government departments naturally were also expected to use Japanese machines. Although many firms and even some government offices lobbied for permission to import IBM computers, MITI only occasionally and reluctantly granted permission. If Japanese users did not use Japanese machines, MITI's argument went, then it would not be possible to convince potential buyers in overseas markets to do so.

The main objective for both the computer companies and the Japanese government was to make Japanese computers which could compete with IBM's latest machines. To reach this goal, the government sponsored a series of government–private sector cooperative research projects. The first of these was called the FONTAC project and it was launched in September 1962. Fujitsu was selected to develop the main processor and NEC and Oki were to make the peripheral equipment. However when the project was completed two years later, the various parts of the machine did not fit together properly and it did not run. Although this first project was generally a failure, the firms learned a great deal which would be valuable in the future. The government was also made clearly aware of the level of financial resources which would be required to develop a Japanese computer industry. This point was heavily underscored as even before FONTAC's completion, IBM had launched an even more sophisticated series of computers.

The Super High Performance Computer Project (1966–72) was the second major cooperative research project. Its goal, which would require the commitment of the entire Japanese computer industry, was to create a machine superior to IBM's latest model. The government committed 12 billion yen to the project, put together a description of the desired computer system to be developed, and asked Hitachi, Fujitsu and NEC to submit proposals. Hitachi's proposal was selected and it became the project leader. Along with Fujitsu and NEC, Hitachi would research mainframes and integrated circuits and in conjunction with the Japan Software Company, would design the project software. Work on the peripheral equipment (e.g. character recognition and display) was assigned to Toshiba, Mitsubishi and Oki Electric. (This division of labour left the latter three companies little opportunity to develop large computers.) Although the Japanese makers did not succeed in making a superior machine to IBM's, they did achieve many of their specific hardware objectives – memory capacity, data transfer speed, addition speed – and in a very short period of time the cooperative project helped the development of the industry dramatically.

In 1968, NTT (Japan Telegraph and Telephone Company) entered the computer arena and announced plans for a cooperative research project to develop a high performance computer. NTT was concerned about the availability of Japanese technology for data communications and services. NTT fell under the authority of the Ministry of Posts and Telecommunications, a ministry often involved in turf wars with MITI, but this time both ministries were focused on nurturing the computer industry and believed that NTT's participation would be a positive step. NTT's first project, DIPS-1 Project, was to take the preliminary results from MITI's Super High Performance Computer Project and attempt to commercialize them into a machine that could be used to offer on-line data communications. While MITI's project had very high goals (a machine better than IBM's), DIPS-1's objective was a machine that could be used to offer data communications as soon as possible. Hitachi, Fujitsu and NEC were involved in this project also and NTT encouraged competition among them and had each one develop its own DIPS-1 machine independently. The first machines were sold to NTT in August of 1972 and were used to support a public data communications system. The Japanese firms achieved most of their technological goals, entered the telecommunications market and gained invaluable experience. Frustratingly for the Japanese companies, however, on the heels of their DIPS-1 launch, IBM announced the completion of its newest and much more sophisticated machine.

Therefore, by the early 1970s despite the efforts of both the government and the computer makers, the Japanese computer industry was facing tough times. Not only was IBM even further ahead of the Japanese companies than ever but Japan was being bombarded with increased calls for liberalization of the market. In the summer of 1971, Japan agreed to liberalize the industry by the end of 1975. The Japanese makers were left with four years in which to become competitive. The government felt that the industry needed to become more consolidated and thereby gain research and development and production economies of scale. MITI began to pressure the computer companies to merge into, ideally, three firms. The companies were very reluctant but eventually agreed to restructure into three groups; each group would develop a different size and style of computer. Fujitsu and Hitachi would develop the large computers; NEC and Toshiba medium-sized ones and Mitsubishi and Oki small computers for specialized uses. This temporary segmentation of the market would, it was hoped, enable the industry as a whole to counter IBM's full line of machines.

This arrangement, referred to as the New Series Project, was given a subsidy of 70 billion yen, and a deadline of four to five years. Not everything proceeded smoothly. Hitachi and Fujitsu were each too large and powerful to want to ever take the backseat to the other which resulted in a difficult working relationship. The Mitsubishi-Oki machines were not a big success. Many technological advances were made, however, and by the mid-1970s, the companies were making a strong comeback against IBM. While some of this success can most definitely be attributed to alliances with foreign firms (Fujitsu's relationship with Amdahl and NEC and Toshiba's alliance with Honeywell), 'the project made the difference between the industry's success and imminent failure'.

As the New Series Project progressed, Fujitsu, NEC and Hitachi were also involved with NTT in its DIPS-11 project. Beginning in August 1973, these three companies were again given assignments to develop computers to different technical requirements but all for use in NTT's telecommunications system. The objective was to upgrade the DIPS-1 machines prior to the liberalization of the market and this was achieved.

MITI also launched another project. This one – the Pattern Information Processing Project (PIPS) – was designed to give the Japanese firms an opportunity to 'gain a competitive advantage in the Japanese market by offering sophisticated Japanese language systems, something IBM and other foreign makers were not investing in'.[6] PIPS's objective was to develop a pattern recognition system, ways for computers to recognize written and vocal patterns. The Japanese companies were not very enthusiastic about MITI's proposal and only agreed to participate when MITI indicated it would foot the entire cost. The projected lasted from 1971 to 1980 and cost 22 billion yen.

The lack of enthusiasm for the project on the side of the companies was due to their real concerns that it would be difficult to achieve any kind of commercial application from the research. In the short term, this proved to be true and the project was initially perceived to be a failure. Although the companies achieved most of their goals–Toshiba was investigating one system to recognize printed characters and another to recognize shades of black and white; Hitachi looked into object recognition, NEC voice pattern recognition, Mitsubishi colours – at the end of the project much still needed to be done to turn the initial research into commercially viable systems. Nonetheless, within the decade many of the systems developed under the PIPS project have resulted in commercially successful products for the companies involved.

MITI's influence over the computer makers gradually declined as the companies became increasingly competitive. In the late 1970s and early 1980s, international pressure for more open markets also meant that the government had to become a little more covert in the assistance it offered. MITI nonetheless remained involved in directing and assisting the Japanese computer industry. While earlier cooperative research projects had catching up to IBM as their primary objective, in the late 1970s, the goal was to create a technological advantage, to move ahead of IBM. MITI, NTT and the Japanese computer firms all agreed that the next research project should be the development of the next generation of computer chips called VLSI. NTT, in fact, had begun its own research into VLSI in 1975 in conjunction with Fujitsu, NEC and Hitachi. The NTT and the three company collaboration developed a number of important technologies and created more and more highly integrated computer chips.

MITI's VLSI project ran from 1976 to 1979, cost 72 billion yen and saw the companies working in two groups. Mitsubishi joined with Hitachi and Fujitsu, Oki was dropped and NEC and Toshiba remained partners. The firms were to research both the more quickly commercializable technology that they wanted to study and the more long term technologies desired by MITI. The selected research topics looked at the technology for developing VLSI chips and the machines to test the reliability of the chips. The VLSI project was a definite success, resulting in 1,000 patents and contributing 'greatly to the Japanese makers' ability to produce low-cost, high quality VLSI in the late 1970s'.[7]

There is little question that the Japanese government's involvement in the development of the nation's computer industry was vital. The cooperative projects allowed the firms to combine their resources, both financial and human, and share the risk which is inevitable in research of this magnitude. Having different groups take different approaches to solving the same problem increased the likelihood of success. The government financing was, of course, also of considerable importance. It would have been very difficult for any of the companies to shoulder the high costs of research and development, particularly at the outset of the development of the industry. There is a reasonably good chance that the Japanese computer industry may have developed without the assistance of MITI and the other government departments but there is no doubt that it would have taken a good many more years to do so.

By the 1990s, Japan, for all its economic challenges, had become a world-leader in computer products and utilization. They did not keep up in terms of Internet use. The United States, in contrast, remained the world leader and, throughout the 1990s, defined the contours of the Internet revolution. Japan lagged well behind (see Table 1.1), purchasing fewer personal computers, creating many fewer web-sites, and largely failing to engage with the new technology. Today, Japan has formidable capabilities in the growing field of computerization, the industrial application of computers, and the development of consumer products that capitalize on digital technologies. In two of the three fields – industrial computing and digital consumer products – Japan is a world leader, and the nation's accomplishments in the general field of computerization are impressive as well. The country's numerous research centres, institutes and product development operations keep Japan at the cutting-edge in the development of computer technologies. Through the 1980s and 1990s, the government's concern about the importance of computer technology to international competitiveness – combined with its continued belief in government-directed economic growth – resulted in numerous government-business initiatives in this sector. These activities ranged from the development of universities specifically designed to provide advanced training in computer technologies to a national commitment to the technopolis concept (a marriage of local governments, universities and corporations around a specific high technology theme). While most of the latter foundered, the commitment to technological development bore fruit overall.

Japan has maintained a competitive advantage in both the manufacture of consumer and industrial products. The fascination of the Japanese consumer with consumer and portable electronic goods created a rich, dynamic local market for personal computer products. The country responded slowly to the availability of the new consumer machines,

Table 1.1 Computer hardware investment, US and Japan[8]

	US	Japan	US/Japan
Total Computer Shipments, 1993 (US$)	62 m	28 m	2.2
PC Installed, 1993	67 m	9.4 m	7.1
PC Shipments, 1994	18.6 m	3 m	6.2
Number of www sites, March 1995	8,800	306	28.8

selling only 2 million computers in 1990 and less than 1.8 million two years later. By 1996, domestic shipments of computers peaked at over 17 million before leveling off the next two years at between 16.3 million and 14.5 million units. On the domestic front, fewer than 20 per cent of Japanese homes had personal computers in 1996, a number that rose to over 57 per cent by 2002. By the mid-1990s, computers had become, in Japan as elsewhere, a significant feature in Japanese homes and businesses – but the expansion was only beginning. Few countries in the world produce the range and number of digitally-based wares on a regular basis – and none find as receptive an audience for the new offerings. At the same time, Japanese industry has shifted its focus to the production of high-end technology, emphasizing items that fill a critical niche in world-wide computer-based markets. The country's firms have largely moved assembly and low-end product manufacturing off-shore, to factories in Thailand, Indonesia, China and the Philippines, retaining the high-end work for the domestic producers. This combination has generated a high technology industry in Japan that is far larger and more diverse than is generally recognized and that ensures the country a continuing and significant place in the global technology marketplace.

Even a quick visit to Japan reveals the popularity of consumer computer-based products. The country's stores are full of digital items, ranging from digital cameras, personal computers, portable computers, computer games, and the like, to a range of products that have, to date, attracted less international attention (home appliances connected to the Internet, robotic toys, and a fascinating array of digital music products, including soundless drum sets and world class high fidelity music centres). The roll out of new items routinely become front-page news. When Sony released its Playstation II in the spring of 2000, Japanese consumers bought up every available machine within a matter of days – more than half a year before the same product arrived in North American stores. The release of new games for Playstation, Sega's Dreamcaster, or Nintendo's system attracts huge crowds of eager teenagers. And each week sees the release of numerous new consumer items, many of which prove less than popular with local consumers and never make it into wide distribution. To fuel interest in their many products, the major manufacturers maintain product information centres in the largest centres. Sony's main display in Tokyo, for example, is a multi-story complex, with each floor dedicated to a specific portion of their product line. Crowds numbering in the thousands pass through the complex every weekend, many stopping in the store at the bottom to buy one of the items they have tried out.

Corporate interest in high technology is no less pronounced. The high cost of Japanese labour, combined with the decline in the workforce and a desire to innovate, has encouraged companies to seek ways to integrate digital operations into their production processes. The result has been the evolution of the world's most advanced factories. In the 1970s and 1980s, Japanese companies attracted a great deal of attention for their production methods and organizational strategies. Just In Time Delivery, Total Quality Management, quality circles and the like provided world-leading examples of how to better integrate workers and suppliers into the manufacturing processes. Through the 1980s and 1990s, however, an ever more dramatic revolution saw a vast expansion in the use of robots within companies, with Japan establishing itself as a world-leader in both the production and use of robots in manufacturing. (There is a joke circulating to the effect that the Japanese manufacturing plant of the future will be staffed by robots, a dog and one person. The worker's job will be to feed the dog which, in turn, will keep people away from the robots!) There is no better symbol of Japan's responsiveness to the commercial potential of information technology and computerization than the impressive robot-based manufacturing plants which produce a significant portion of the country's industrial output. In this area, the country has no substantial rival.

While consumers and observers marvel at the latest products from Sega, Nintendo, Sony, Hitachi, and other Japanese firms, very little is known about the industrial robots which underlie much of Japanese industry's success. Japan has, over the last forty years, become known for the consistently highly quality of its products, a marked departure from the image of shoddy workmanship and uneven production standards from the 1950s and 1960s. As a significant part of the shift toward an emphasis on quality, Japanese firms began to work with industrial robots. While the first machines were heavy, cumbersome items, best suited for large scale applications, continued refinements resulted in the production of increasingly specialized and efficient industrial robots. Between the 1960s and the 1990s, industrial robots replaced tens of thousands of workers in automobile, television, computer, and other manufacturing plants. (Japanese workers did not, in the main, rebel against the introduction of labour-saving devices. They were protected by the growth, through much of this period, in industrial output and by the lifelong employment practices of the major corporation. Thus shielded from losing their jobs, the workers placed the interests of their employer above the protection of their specific tasks.) Japan's advantage in robotics is considerable. By 2002, the country is expected to have close to 370,000 industrial robots in operation. The United States, by way of comparison, will have approximately one third that number and the United Kingdom lags far behind, with only 15,000.[9] The future is not unrelentingly optimistic, however. The availability of cheap labour in Asia has undercut some of the domestic demand for factory robots, leading to declining interest in the production of these critical industrial tools.[10]

In the first years of the 21st century, Japanese firms once again led the way with robotics, this time with the development of robotic 'pets' for eager Japanese consumers. Sony's *Aibo*, which was launched in late 1999, was the first and despite its high price tag – about US$1,500 – was immensely popular. *Aibo II* was released in October 2000 and has extra features including being able to read out e-mails and take photos.[11] Sony launched a new generation of the *Aibo* family in November 2001. The ERS-220 looks more futuristic than the previous Aibos and can be linked to a personal computer via a wireless local area network card.[12] In October 2001, the toy company, Tomy, released a communication robot, the *Memoni*, which comes with a 20,000 word vocabulary and can learn another 3,000. At the same time, Omron began marketing a robotic cat with synthetic fur called *NeCoRo* (neko means cat in Japanese). *NeCoRo* cannot walk but it can turn and look or reach out to things and can create personalized behaviour for different owners.[13]

Japanese newspapers and magazines are filled with announcements of new robots able to walk upstairs, pour glasses of wine, write on a whiteboard, play air hockey (tracking a puck travelling at 30 kilometres an hour and pinpointing it to within 1 cm of its actual position),[14] or navigate a maze. The commercial potential of these robotic abilities are not entirely clear but the developers hope to be able to use them for everything from 'seeing eye' robots for the blind to personal robots which will fetch, clean and generally look after their owners. Toshiba and Sweden's Electolux have developed a self propelled vacuum robot and in April 2003 Tmsuk Co. launched a home security robot named *Banryu*. *Banryu* has various sensors (fire, smoke) and can maneuver even in cluttered spaces.[15] Kawada Industries and the National Institute of Advanced Industrial Science and Technology unveiled their Humanoid Robotics Project humanoid robot in December 2002. This robot can walk for two hours while carrying loads of as much as 6 kg.[16] The Kanagawa Institute of Technology has developed a 'Nurse

Assistant Power Suit' which gives the wearer the ability to lift heavier weights.[17]

Given the Japanese enthusiasm, both corporate and consumer, for robots, it is no surprise that the largest robotics competitions in the world are held in Japan. The 13th Robot Sumo competition, held in 2001, had over 4,000 entries. Scientists who designed robots for RoboCup 2002 – the robot version of World Cup soccer – have set themselves the surreal target of creating a team of robots capable of defeating a human team. They intend to achieve this goal by 2050. The ability of Japanese manu-facturers to combine the Internet, sensors and computer technologies have left the country with a significant lead in the development of robots for consumer and low intensity business use.[18]

Beginning in the 1980s and accelerating rapidly thereafter, Japanese firms became particularly adept at taking the new digital technologies and applying them to consumer products. Many of the world's most popular computer applications – games systems by Sega, Nintendo and Sony's Playstation – originated in Japan, and the country likewise con-tributed significantly to the development of game software and design. The rapid introduction of digital cameras, in ever-growing cycles of innovation and sophistication, gave the country an international lead in this critical market. Japanese companies pioneered the commercial application of digital photography, introduced a variety of memory systems (including storage on floppy disks, memory sticks and, in 2000, cameras which took digital pictures and provided the option of saving the photo on regular film). They pushed the envelope on portable radios and television, including James Bond-like wristwatch televisions, and introduced digital systems into a variety of entertain-ment and home computer devices.[19]

Japanese companies and consumers also were world leaders in cellu-lar phone technologies, which seemed particularly well-suited for the fast-paced, highly mobile domestic market. Cell phones quickly became ubiquitous in Tokyo, Osaka and Hiroshima. They soon emerged as a fashion accessory among the young and stylish. Deregulation in 1994 brought new competitors into the field, with four major firms competing for consumers in the major urban markets by the following year.[20] The price declines which followed only spurred both market demand and product innovations. Japanese companies launched a new system, personal handyphones, using cheaper delivery methods (line of sight, based on numerous towers throughout the city, rather than the standard cellular technology) and targeting their prod-ucts at the highly imaginative youth market.[21] Within only a few years, millions of teenagers had the handyphones, using them to stay in seemingly constant touch with friends and family members. With the advent of i-mode technology, however, the handyphones were 'consigned by some to the dustbin of history', although similar systems were being introduced as a cheap alternative to land lines and cellular phones is such countries as China, Thailand and Taiwan.[22]

Innovations in applications, of course, often have a direct effect on further industrial developments. DoCoMo's unprecedented success in marketing the i-mode phone led the company and its competitors to immediately expand their efforts to improve Internet speeds, service reliability, and other technological aspects of their service. At stake was a huge and growing Japanese market for digital phones and supporting devices and a potentially enormous (particularly in Asia) demand for manufactured goods relating to the i-mode phone. Kei-ichi Enoki, managing director of NTT DoCoMo in Tokyo, exhorted Japanese manufacturers not to rest on the laurels of past success. Even as the company was unfolding its latest innovation – high-speed Internet access over the i-mode phone – Enoki called on companies to hasten development of new computer chips which would permit even greater improvements in the mobile Internet system.[23] (As an important benefit associated with rising to the forefront of its field, DoCoMo has also emerged as one of the preferred employers among university graduates. In the words of Osaka-based analyst Paul Scott, DoCoMo has 'product charisma' and is reaping many benefits from this status.)

Japan has made formidable advances at the interface of basic science and industrial applications, and melds this with perhaps the most innovative consumer manufacturing environment in the world. Japanese firms experiment widely, and produce thousands of products which never make an appearance outside the national market. They are also confident about their technological abilities and commercial acumen. (Sony's Chairman Nobuyuki Idei, early in 2001, asserted that the US was over confident about its broadband communications systems and would face a major challenge from Japan in this field.[24]) The nation's scientists, in research laboratories, government-business consortia, and academic-industrial cooperative projects are linked to major research initiatives the world over. Aided by massive infusions of government money over the past two decades, these research initiatives are, in financial and technological terms, the envy of international collaborators and the locus for an enormous amount of cutting-edge work. At the corporate level, investment in research and development remains extremely high, in comparative terms, and is further fuelled by institutionalized cultures of innovation in product design, manufacturing processes, and responsiveness to consumer tastes.

Many observers have underplayed Japan's continued high-technology accomplishments, but competitors underestimate the country at their economic peril. As Eamonn Fingleton pointed out in *Blindside: Why Japan Is still on track to overtake the U.S. by the year 2000,* Japan's failure to dominate certain high-technology sectors has masked its formidable general accomplishments. Western criticism of Japanese advances peaked in the midst of the dot-com boom of the late 1990s, when the country's advances in software design and application lagged well behind North American and European developments. The Japanese fascination with digitally-based products and computer-oriented technologies is itself a focus for considerable industrial development and innovation. Japanese firms played a critical role in the development and commercialization of such widespread technologies as the photocopier, digital camera, and portable computer. Observers, however, also failed to check inside the machinery that was driving the digital economy. Had they done so, they would have noticed some critical developments. Their emphasis has focused further on the production of the most critical and valuable components of these items. For example, Japanese firms dominated the global market for LCD (liquid crystal displays) for portable computers for several years, ensuring that Japanese technology showed up in virtually all portable computers, regardless of nation of manufacture. (The Japanese lead in this crucial area has recently passed to Taiwanese and South Korean firms, which are developing a new organic electroluminescence display panels.)

Close business cooperation through the highly interconnected *keiretsu* and strong government encouragement resulted in the development of price competitive and quality leading digital technologies in a variety of sectors. Major firms, including NEC and Fujitsu, emerged at the forefront of international trade in such key areas as DRAM (dynamic random access memory) chips and digital storage systems. Japanese firms contributed and marketed critical conceptual work in such fields as fuzzy logic and neural networks. Technical people, much more so than business analysts and government officials, recognized the significance of Japan's corporate investments and successes, and urged national competitors to pay more attention to Japan's significant advances.[25]

Yet, in technology, as in so much else, Japan is a land of contradictions. As has been discussed, the country is famous the world over for its high technology industries and advanced industrial techniques. However, on other fronts, Japan lagged well behind. For example, until the appearance of DoCoMo and the i-mode phone, digitally delivered services had made little impact in Japan. Bank machines that had become commonplace in other countries were difficult to find in Japan. Foreigners would search high and low in major cities for one of a handful of international cash machines. Retail and grocery stores do not provide direct debit payment; the use of cash remains widespread in the country. Telephone bill payment and automated ordering machines – the bane of life in many western countries, where it has become rare to speak to a human operator – have made very few inroads in Japan. Where other countries appear determined to eliminate staff-intensive payment systems in stores, movie theatres and the like, Japan appears content to maintain them. (The one long-standing exception – a matter of logistical necessity more than cultural choice – is the train and subway ticket systems, the vast majority of which are automated and extremely efficient.) The main reason for this difference is the Japanese approach to service. Face-to-face service, attention to detail, a preoccupation with packaging, and similar customer service traits dominate Japanese business culture. For consumers and managers raised in this environment, the prospect of shifting to an impersonal electronic system has long been resisted.

There is, however, clear evidence that a major shift is now underway. Convenience stores have become a centre-piece of Japanese e-commerce, providing a wide range of digital services. The country does not have effective Internet-based payment systems, but millions of consumers pay their bills at nearby convenience stores. Similarly, the country's major ticket sellers – Pia, Lawson and E-Plus(Sony) – work through the same store networks, which host easy-to-use kiosks for the purchase of tickets to sporting and cultural events. (Unlike their western counterparts, these kiosks do not accept direct credit card payments. Reservations are confirmed through touch-screen technology. Scan-able receipts are then produced which counter staff convert into ticket vouchers. The required intervention by clerks undercuts the efficiency of the system somewhat.) Several major e-commerce companies are working with convenience stores to develop delivery systems for Internet orders. These changes, added to the service provided by the Internet telephones, portend a radical shift in direction for the Japanese retail and service sector.

Despite marked success in other areas of digitized entertainment – most notably *anime* – Japanese television is decidedly low tech. Game shows operate on 1950s-style North American stages, with few of the bells and whistles that dominate American productions. Television news is particularly intriguing in this regard. News hosts often hold up hand-held graphics to illustrate their stories, or use a pointer and wall-board to provide detailed information. Production values are mediocre by western standards, and there is little evidence of the availability and use of advanced digital technologies and computerized production techniques. The gap between western and Japanese television production is likewise evident in televised sports, which are low-key, offering limited camera angles and few gimmicks from the announcing booth.

The issue with television is more complicated than this summary suggests. Japan has the technological capacity to do a great deal more. In fact, many of the key components in television production in other countries are made in Japan. Sony, Hitachi, Matsushita and others are world-leaders in digital imaging and produce many of the world's editing suites. It is puzzling, therefore, that this technology is not used to full effect. The television companies have substantial revenues (and a captive market, as cable services have made few inroads in the country) and very large audiences. It is choice, rather than resources or ability, which results in the low-key, low-tech television programming. That decision, in turns, appears to rest on the Japanese respect for simplicity in the domestic sphere; the country does not demand American-style reporting of baseball or soccer, is not overly impressed with glitzy high-production value game shows, and is comfortable with the homey, accessible 'feel' of the vast majority of the mainstream television programming.

Television shows are but one example of the manner in which Japan has resisted high-technology changes within the country. Expensive high-tech buildings dot the urban landscape, but retail stores, restaurants and other services provide little evidence of a digitized society. Advertisers do no bombard viewers with sales pitches for computers and digital services; most of the high-tech products are sold in specialized stores or shopping districts and through direct, street-side pitches to consumers. Superficially, then, the culture of digitization does not appear to be deeply entrenched, with most of the crucial applications of technology – such as industrial robots – based outside of the public's view.

It was, however, in embracing the internet that Japan particularly fell behind. By 1997/1998, between 8 and 9 million Japanese people had access to the Internet. The number skyrocketed to almost 50 million by 2002, a formidable advance by almost any standard (see table 1.2). In

Table 1.2   Japanese Internet users, 1997–2002

Date	Number of users (m)	Source
02/2002	49.7	NetRatings Japan
12/2000	47.0	ITU
03/2000	21.21	NetRatings Japan
04/1999	18.0	Asia Biz Tech
10/1998	8.8	Access Media International
09/1997	8.0	Dataquest

general, however, Japan lagged approximately three to four years behind competitor nations in signing onto the Internet. Consider the response in the corporate world. In 1995, fewer than 12 per cent of Japanese firms were connected to the Internet. That number grew by the following year to half of all firms and to 80 per cent by 1998. Similar patterns emerged in terms of internet hosts, with the number of Internet hosts increasing 80 times between 1994 and 1999 and the number of Japanese domains soaring from 1,500 in 1994 to over 68,000 by 1999.[26] For reasons that are not yet clear but that may be culture-bound, the Japanese government was slow to realize there was anything of value to be gained from the internet. Numerous western commentators also blamed the slow response on an old canard: an educational system that discourages creativity.[27] Despite the enormous corporate and government effort which continues to be focused on the goal of ensuring that Japan remains a world leader in the production of computer-based technologies, it is extremely ironic that Japan, responsible for a significant portion of the technology underlying the digital revolution, was much slower than other nations in responding to the opportunities presented by the IT transformation, particularly as it relates to the Internet.

Internet charges for commercial and personal use remain, by international standards, extremely high.[28] Japan's much-debated process of deregulation did not proceed at Internet speed. The government relaxed telecommunications regulations in a slow, incremental fashion through the 1990s, but the changes were not fast or sufficiently dramatic to overcome the country's deficiencies in this regard. Japan is alone among the leading industrialized nations in the world in having a 21st century industry and a 1970s telecommunications regulatory framework and infrastructure. Even more importantly, Japan's Internet systems have not yet found the correct price point. Regular and broadband services cost a great deal more than in most industrialized nations, limiting consumer and business use and restricting the adoption of the new technologies.[29] Until costs are brought under control, through deregulation and greater competition, Japan will fall far short of capitalizing on the potential of the digital revolution.[30]

Beginning in 2000, Japan began to catch up in providing broadband services. From well under one million subscribers to broadband services in March 2000, the country had over 6.1 million households connected by September 2002. Government estimates suggest that close to half of all households, almost 20 million in total, will have broadband connections by 2005. The infrastructure development, albeit belated, has sparked a rapid increase in related content businesses, including on-line video gaming, and pay-per view videos. Bandai, a major Japanese animation firm, provided access to its popular *Gundam* cartoons via video-streaming as it tested the market for fee-paying online services. Fuji Television, Tokyo Broadcasting and Asahi National Broadcasting share a single site (Tresola Corp) where subscribers (1,000 yen per month) can view old television programs. Toei uses the same technology to provide movies on line. Companies have been monitoring and surveying the emerging market with considerable interest. Several key firms, Matsushita and Sony, have developed devices which permit broadband content to be viewed via a television instead of a computer screen.[31]

Given the rapid pace of technological development and the late entry into the digital race, it is hardly surprising that Japan faces a significant shortfall in the number of available computer scientists and technicians – hardly a unique challenge – and is attempting to mobilize its school and university systems to meet the demand (as well as, for Japan, taking the revolutionary step of further liberalizing its immigration system to attract more foreign-born technical workers). The country had, however, moved slowly in building IT capacity in the school system. In 1990, only 40 per cent of elementary schools and three-quarters of junior high schools and senior high schools had installed computers. By 1998, almost all of the country's schools had computing facilities.[32] A 1999 survey revealed that two-thirds of senior high teachers could operate a computer; slightly more than a quarter were sufficiently computer literate to offer instruction on computers.[33] The universities have not moved into the field with sufficient alacrity to overcome the nation-wide shortage and are working with IT firms to prepare more students for the field.[34] As the country well knows, past accomplishments are no assurance of continued success in highly competitive, technological areas such as computer applications – and now Japan, ironically, faces the reality of international rivals applying practices of industrial imitation on Japan's industrial and consumer output. This is not a field for complacency.

Japan has also developed a remarkable capacity for integrating the Internet into national life ways. The development of the *keitai* (Internet delivered through a portable phone) is but one example of the manner in which Japanese companies and product developers have sought to adapt Internet technology to the unique circumstances facing the Japanese people (see Chapter 3 for an extended discussion of the *keitai* revolution). There has been a dramatic expansion in wireless Local Area Networks (LAN), positioned within coffee shops, libraries, fast food stores, airports, community centres and shopping areas to attract portable computer users and, in a 2002 experiment developed in line with the soccer World Cup, introduced into first class sections on East Japan Railway trains. Mos Burger, one of Japan's largest fast food companies, introduced LANs into a selection of its stores in a bid to attract patrons[35]. (It is worth noting that Japan produces and sells dozens of models of mini-computers, approximately one-quarter to one-third the size of a standard portable computer and particularly well-suited to wireless operations.)

Japan finds itself at a critical cross-roads.[36] The country was not, in the early years of the 21st century, the most wired, computer-ready, digitally-advantaged nation in the world. [37] The Mori and Koizumi governments have been promoting the Internet and digitization with the unreserved passion of the latter-day convert, but formidable barriers remain, particularly those related to a complex, expensive and over-regulated telecommunications system. While, as DoCoMo has proven, this need not be a fatal flaw in the Japanese system, the country is not well-prepared – on the surface –to embrace the digital revolution. Classrooms lag behind those in most western countries and small business and government are only slowly going on-line. Despite the belated start, however, Japan possesses some critical advantages, ones which may well propel the country to the forefront of the digital age. Japanese consumers give innovative products a chance at success. The country is full of what marketers call 'early adopters,' consumers who are willing to experiment with new products. The major companies conduct the normal market tests and reviews, release new items into the marketplace and then decide if they are going to keep them in production, revamp the design or concept, or test the products in foreign markets. Japan offers manufacturers a wealthy, fast-moving, and highly forgiving test market, and many of the leading IT firms have capitalized on this important commercial foundation. The remarkable success of the Walkman, Playstation, DoCoMo mobile Internet phones, Sega systems, and other digital devices are a testament to the willingness of Japanese consumers to try out new items and to shift to them *en masse* if the quality, price and nature of the product warrants.

Japan's experience suggests that values, social relations, consumer movements, and cultural patterns have both constrained and stimulated the expansion of digital delivery systems. Technology is both culture-bound, in that it reflects the values of the production system from which it emerged, and culturally framed. Not all societies, certainly not Japan, will respond to new products and services in quite the same way. Those technologies which are well-suited for the culture, lifestyles, and economic needs of a society will be adopted quickly; those which do not fit with local realities or which threaten well-entrenched social relationships or understandings are likely to do less well. There is, to put it simply, no simple pathway through the digital revolution, and Japan's approach to the threats and promise of the information age will invariably reflect the country's unique circumstances.

Industrial and commercial developments from the 1960s to the first years of the 21st century created a solid foundation for Japan's digital economy and Information Technology society. When the Internet emerged, later in Japan than other countries, the capacity for isolated, independent computers expanded exponentially, as the capacity for networking and digital communications became quickly evident. The advent of the World Wide Web and related communications systems popularized computers in ways hitherto unimagined, creating the potential to recast economies, restructure businesses, reorient personal and social relations, and revolutionize interpersonal and mass communications. But the Internet required a technological and production base. It needed, and needs, companies producing consumer and commercial applications and devices, governments pushing the need for people to be 'connected', service industries creating content and reasons to use the new technologies, and a society willing to experiment with the services and equipment of the digital age. Japan may not have been as ready as other countries for the emergence of the Internet as a major economic and social force, but its industrial and commercial sectors were, in particular, eager to capitalize on the promising opportunities of the new economy.[38]

Public, administrative and commercial acceptance of new technologies is never assured, and there are myriad factors which influence a nation's ability and willingness to capitalize on innovations and industrial developments. Japan raced ahead of its competitors on some aspects of the digital revolution – industrial robotics and mobile Internet devices being among the best examples – and lagged woefully behind on others, including general Internet use and the development of competitive e-commerce models. If, as promoters of the 'new economy' long argued, national economic success hinged on a country's ability to respond to the challenge of computerization, Japan's record has been decidedly mixed. With the euphoria of the Internet investment bubble exploded by the dot.com meltdown of 2000/2001, Japan's comparative inability to capitalize on the commercial 'promise' of the e-commerce revolution can now been seen in a more respectful light. But as those who follow the digital transformation have long observed, retail trade over the Internet is only one small portion of a technological restructuring with widespread ramifications for the entire world. How Japan responds to the opportunities and threats presented by digitization will, as for all nations, have a strong influence on how the country develops economically and socially in the decades to come.

2
Japan.com: Government and the Promise of the Internet Society

As the Internet developed, emerging out of the uniquely scientific and strategic communications environments of university research laboratories and the military, it was not altogether clear that this new technology had a great deal of potential. Governments generally viewed the Internet as an adjunct of the scientific enterprise; very few people saw much general or public use. By the early 1990s, that had changed. The Internet spread beyond and through university campuses. Government departments discovered the potential for internal communications. And, most critically, the private sector came to the realization that this odd combination of computer and telecommunications technologies might well have commercial applications. Governments had the potential to speed up the development of the Internet, by investing in technological infrastructure, training and scientific development. They also all had regulations, procedures, licensing requirements and the like which had not anticipated the development of the Internet. By the late 1990s, it had become increasingly clear that the interface of government and technology would play a critical role in determining the role and impact of the Internet within specific societies. What would be determined on a country by country basis was the degree to which the government's role would be constructive or constraining.

While Japan was slow to recognize the commercial importance of the Internet, the country's fascination with digital equipment put it at the forefront of the consumer and industrial computing industries, a lead that was declining by the mid-1990s.[1] As computerization took hold around the world, and as it expanded into such fields as mobile telephony, the Internet, business to business e-commerce, enterprise management software, robotics, and the like, Japan joined many other industrialized nations in believing that technological leadership was critical to long-term national success. Japan's sense of urgency was increased by a growing awareness of its own changing demographics, the most important elements of which were a rapidly aging population and a forecast steady decline in population in the 21st century. Faced with the realization of workforce shortages and the growing needs of the country's burgeoning group of seniors, the Japanese government looked for technologically mediated solutions. Computerized systems promised to replace workers, increase efficiency and productivity, maintain competitiveness and ensure that the remaining Japanese workers had an opportunity to earn high personal incomes. To this end, and with increasing intensiveness over time, the government of Japan and its corporate partners sought cooperative, collaborative approaches which had the potential to keep the country at the economic forefront while maintaining opportunities for Japanese workers and businesses.

In the mid-1980s, faced with the increased pace of technological change and the domestic problem of regional economic inequality, the government embraced the concept of developing technology centres. The technopolis, given legislative shape under the Technopolis Act of 1983, sought to draw together the national and prefectural governments, businesses, and the academic sector. Each technopolis was intended to establish a world-class capacity in a specific industrial or technological sector. Governments provided land, infrastructure support and start-up funding. Companies offered investment capital, industrial expertise, and manufacturing and marketing ability. High schools, colleges and universities were mobilized to provide a highly trained workforce, well-prepared for the specific technological challenges at hand.

The goal of the technopolis initiative was to broaden economic opportunity outside the Tokyo–Osaka corridor and to ensure Japan's continued technological leadership. Each of the 26 technopolis had a different focus. The Yamagata technopolis (Yamagata prefecture is in northeastern Honshu) focused on life support technology, including such diverse activities as artificial organs and preventative medicine for aging. Hamamatsu (located in central Honshu) focused on optical technologies. Several of the sites emphasized more traditional technologies, albeit with a cutting-edge twist, like deep sea mining. While few of the specific initiatives highlighted IT-related products, a number, like the Kumamoto technopolis (in central Kyushu) emphasized information services and software. Although not called a technopolis, the Gifu Prefectural Research Institute of Manufacturing Information

Technology opened in early 1999 and supported research in robotics, computer simulation, virtual reality and computer vision.

While there were some specific successes, the technopolis concept fell far short of its goal. The combination of political agendas, local realities, and financial difficulties at the national level undercut the technopolis initiative. Creating industrial and digital powerhouses outside the metropolitan centres – a Japanese variant of an international effort to clone the Silicon Valley achievement in other locations –proved a formidable challenge. Hokkaido's Sapporo Valley, which developed at least in part because of a small group of computer researchers from Hokkaido University, is one of the most successful attempts. With a concentration of research facilities, Sapporo Valley 'has fostered about 300 companies, including some that have produced original world-class technologies'.[2] Certainly some of the credit for this goes to the municipal government which provided strong support to IT start-ups, including launching various pilot projects and providing facilities for software developers to test Java-based programs. Even the expenditure of significant amounts of government funding could not overcome logistical difficulties, the shortage of creative ideas, or the forces of international competition. By the late 1990s, the early excitement generated by the technopolis concept had waned significantly and most of the specific plans had been abandoned in favour of more generic efforts at industry-government-academic cooperation.

Colleges and universities tied to the needs of specific industrial or technical sectors sprang up around the country. In 1987, the Japanese Ministry of Education began establishing centres for cooperative research at national universities. As of April 2000, there were 53 centres.[3] Local governments bought into the concept of major investments in industrial infrastructure and research and development; sizable collaborative projects, most freed from the tethers of the theme-specific technopolis idea, survived. And while university–business cooperation in Japan works very differently than the intense interaction that has come to characterize business development activity in North America and Europe, the universities responded to the increased emphasis on technological development. Between 1987 and 1997, the number of joint research projects between national universities and the private sector grew by six times.[4] In April 2001, the University of Tokyo established the Advanced Science and Technology Enterprise Corporation (ASTEC). ASTEC was designed to send researchers to companies to work on the development of prototypes. At the same time,

Kyoto University signed an agreement with Rohm Co. to begin joint research on a new generation of semi-conductors and another agreement with Sharp Corp to research nanotechnology.[5]

On a broad national scale, there remained a significant commitment to state investment in research activity, prefectural involvement in university–business cooperation, and coordinated government involvement in industrial and technological development. The almost magical results of the 1960s were not quickly replicated, nor was the success of Computers Inc. soon evident in the Information Technology sector. But the idea of multiple business, government and education partners working together to further the country's economic interests remained firmly entrenched. If anything, Japan did not learn enough from the shortcomings of the technopolis concept and remained more committed to the idea of centrally led innovation than, as western critics repeatedly observed, to liberalizing the corporate environment as a spark to innovation.

There are numerous examples in Japan of the continued importance of large scale cooperative developments in the technological field. Two of the most impressive are the Tsukuba Science City, located in the Tokyo region, and Kansai Science City, constructed between Osaka and Kyoto. Tsukuba Science City was first envisioned in 1961 although it was many years before it came into being. Tsukuba has 'about 300 major research, education and scientific exchange facilities'.[6] The Kansai Science City project, launched with the passage of the Kansai Science City Construction Act in 1987, sought to bring educational institutions, business and government agencies together in a large science-related complex located in the country's second largest economic centre. This model city endeavoured to promote research on new technologies and to thereby spur economic activity in the region. The initiative 'is also considered as one of Japan's national projects aimed at constructing a stronghold for the development and production of creative, international, interdisciplinary and cross-industrial culture, science and technology for the 21st century, and at contributing to the development of culture, science and technology for the Kinki area, as well as of Japan and the world'.[7]

The Kansai Science City plan was ambitious, aimed at creating a research and development complex which spanned a large area, drew on resources from existing and newly built institutions and agencies, and stimulated economic activity as a result of scientific and technological investigations. Promoters set a population goal of over 400,000 and plan for over 80 separate research facilities. Work began apace, fuelled by the residual revenues of the bubble economy and the pump-priming enthusiasm of a national government attempting to reverse the economic decline. A partial list of the facilities integrated into the Science City gives an indication of its diversity: Advanced Telecommunications Research Institute International, Ion Engineering Research Institute, Nara Institute of Science and Technology, Research Institute of Innovative Technology for the Earth, a research initiative of the Foundation for Multimedia Communications, Nara Research Center of the Telecommunications Advancement Organization of Japan, Tsuda Science Core, and the Advanced Photon Research Centre. Japan had, in its earlier work on automobiles, industrial technologies (including robots), and the computer industry, invested a great deal in the belief that juxtaposing industry, government and education would stimulate economic growth. In the process – as is hoped for Kansai Science City – these cooperative initiatives created world-class research centres, capitalizing on unique opportunities for synergy and interaction.

In 2001, one of the most interesting government-sponsored projects was launched. Japanese researchers from academia and the corporate sector (including Toyota Motor Corp, NEC Corp., and Denso Corp) have begun work on the development of an 'internet car'.

The Ministry of Economy, Trade and Industry (METI) contributed half of the estimated Y2 billion in expenses. Plans call for the car to be capable of providing the driver with access to the Internet which will enable him or her to enjoy a wide range of music and video entertainment, make hotel and restaurant reservations, pay tolls automatically, receive information on traffic conditions, and collect detailed weather information on the area in which it is traveling. This will be the first experiment in applying Internet Protocol Version 6, a next-generation Internet technology, to vehicles.[8]

Given Japan's level of technological development and the various governmental attempts to promote technological development, it is surprising that the Internet got off to such an inauspicious start in Japan.[9] The Japanese government was clearly slow to appreciate the opportunities presented by the Internet or, more precisely, placed greater priority on maintaining its highly regulated telecommunications system than on exploring the potential of the new technology.

By way of background, the first Internet transmissions from Japan were sent by American engineers working for two US companies,

Intercon Systems and AT&T. Intercon was providing a service for its 400 subscribers, most of whom were expatriates living in Japan. A small group of Japan computer pioneers had hoped to beat InterCon into the field, but their way had been blocked by the Japanese government. Despite months of lobbying, the Ministry of Posts and Telecommunications had refused to grant the Japanese company the necessary license to launch an international Internet system.[10] The struggle persisted in subsequent months, when the Japanese government cut off the e-mail connection operated by TWICS (InterCon's first customer). Japan's growing computer science community resented the government's interference and protested vigorously, but to no avail. As a result, few Japanese were able to capitalize on the emerging technologies.[11] System usage was, in 1993, minuscule, with the Japanese logging only 42,000 megabytes of inbound traffic per month; Japan had less than 5 per cent the number of network connections of the United States. At least part of the reason for this was, as Shigeru Nakayama of Kanagawa University pointed out, because:

> the alphabet based keyboard was a bottleneck for Japanese seeking access to the internet. Indeed, it was a problem for all East Asians who use ideograms, as they get used to alphabet much later than their counterparts in European countries. According to a UNESCO survey of computer literacy in the late 1990s, the Japanese people are far behind Europeans or Americans, simply because the average Japanese person is ignorant of the alphabet based keyboard up until high school time.[12]

As Nakayama went on to explain, while Westerners moved from writing to the typewriter to the computer keyboard, East Asians went directly from handwritten to electronic communications, which proved to be a much more difficult transformation.

Jun Murai, alternately referred to as the 'Godfather of the Japanese Internet' the 'Internet samurai' or as a technological 'guerilla', battled with government officials over the right to bring the Internet into the country. The Tokyo-born Murai studied at the University of Keio, where he developed his skills as the country's foremost advocate of the Internet and, between 1979 and 1984, constructed Keio Science and Technology Network. Keio S&TNet was Japan's first campus-wide local area network. Murai resented the government's iron grip on the use of the telecommunications grid and supported a more activist and demographic approach to the technological developments. As he observed several years after the Japanese Internet was operational:

> Even with technological breakthroughs, humans should decide which direction technology ought to go. We should not rely only on technology. It can be said that technology is similar to vehicles that humans can ride in, but technology does not show us where to go. Communications technology, in particular, is like that. Only after we get in the vehicle can we know how we feel and how far we can go. In this sense, the key issue in creating a new information infrastructure is not only to develop a fast, high-quality multimedia network but also to provide an environment that anybody can use anywhere. The Internet takes the role of the latter, and some parts have actually been realized in current society.[13]

Faced with continuing opposition to the development of the Internet, he and his colleagues proceeded on their own. The first iteration of the Japanese Internet was a small initiative called JUNET (Japan University Unix Network), developed between 1984 and 1994. The system offered a dial-up service using regular public telephone systems targeted primarily at the research communities at several Japanese universities. Language, however, left Japan isolated: 'The Japanese language and the prevalence of character codes incompatible with international standards results in a new kind of natural seclusion today for the individuals on the pasocom tsuusin (personal computer communications) islands of Japan'.[14] The popularity of JUNET led, in 1987/88, to WIDE (Widely Interconnected Distributed Environment), a more substantial electronic backbone supported by several major companies, including Sony and Canon. WIDE also provided a crucial link between Japanese research centres, universities and companies and the rapidly expanding American network. Murai remained active in the promotion of subsequent Internet developments, including the Electronic Observation System (1994), a satellite-based Internet system (1994), the Multimedia Internet (1995), and the Internet 1996 World Exposition, which helped popularize the Internet in Japan. Beginning in the late 1997, Murai's work expanded into the crucial field of mobile and ubiquitous computing.[15]

As happened elsewhere with the Internet, non-scientific and non-commercial users quickly recognized the potential of the new technology.[16] The Japanese government, however, permitted only a very narrowly defined use of the Internet and did not support efforts to broaden access and usage. Murai and his colleagues resisted the government's attempts to limit the use of the Internet and, in 1992, created Internet Initiative Japan (IIJ).[17] This system ran afoul of the Ministry of Posts & Telecommunications' rigid regulations by offering a competing system to the slower, less efficient, and less contemporary operation of the government-sponsored National Centre for Science Information Systems (the government sponsored research organization which had been charged with developing Japan's Internet capabilities.) NACSIS, run by Shoichiro Asano, resented Murai's initiative and saw it as interfering with government plans for the orderly development of the Internet. IIJ had difficultly getting a licence to operate, and struggled to lead Japan into the global Internet world. Murai, himself, was more anxious to return to his research than fight with government, but he and his colleagues persisted. One of the great strengths and weaknesses of the Internet has long been its anarchistic elements, for the technology makes it difficult to monitor usage and access. Murai's efforts prevailed and, over a relatively short period, rendered the government's attempts to control the Japanese Internet ineffectual.[18]

The country's telecommunications structure remained a serious impediment to the development of the Internet in Japan. The dominance of Nippon Telephone and Telegraph (NTT) at one end, seemed to block out potential competitors, while the proliferation of several thousand Internet Service Providers, most offering Internet access within a fairly narrow geographic range, resulted in a highly fragmented, second-tier service industry. The ISPs lacked the economies of scale necessary to secure premium band-width, resulting in much slower and poorer connections at much higher costs than were the norm in other major industrial nations. Gradually, over the years, private ISPs obtained access to better international connections, expanded throughout Japan (primarily by buying smaller companies) and improved service significantly.[19] Prices fell but not as dramatically as in the US and other key markets. As one journalist commented in 1994, 'while Japan may be the world's second-largest economic power, the Japanese remain dangerously isolated. Networking has the power to change that by bringing Japan closer to the international community. But by the same token, failure to log on to the world's largest network could leave Japan more isolated than ever'.[20] At that time, Japan had one-twentieth as many networks as the United States, and its total outbound Internet traffic was approximately that of Taiwan. Australia used the Internet twice as much as Japan.

A combination of factors – high telephone usage rates, the dominance of English on the web, the consequent limited number of Japanese language web-sites, browers unable to handle Japanese, corporate reluctance to grab onto the Internet as a means of advertising, and the slow spread of computers for home use conspired to limit Internet traffic. It is hardly surprising, therefore, that through the mid- and even late 1990s, Internet usage lagged behind that in many other industrialized countries. The January 1995 Kobe earthquake was at least partially responsible for convincing government officials and the public of the considerable benefits of the new technology as the Internet served as a vital means of sharing information at a time when other communications systems proved inadequate. (Between 1995 and 1996, usage in Japan expanded by 41 per cent, by far the highest increase in the world in that time.[21] Asia Pacific went up by 26 per cent and Western Europe by only 15 per cent in that year.)[22]

Convincing government officials of the importance of the Internet was critical. As Canadian Roger Boisvert, founder of Global On-line, Japan's first high quality privately run Internet Service Provider, explained it was important to show why the internet was important.[23] He had to do so to make his proposed launch of an Internet Service Provider attractive to the Japanese bureaucracy. He said, 'I had to present it in a way that would make the government want to do it for me. I had to think of what was in it for Japan. I told them that if Japanese companies do not have access to the most up to date information in the world, then they will fall behind and the country will fall behind'.[24] The government bought the idea sufficiently to give Boisvert permission to proceed. The general telecommunications market followed a similar pattern. In 1995, only two foreign telecommunications companies held the key Type 1 licenses. By 1999, that number had risen to 20. NTT's rates fell in line with the competition. By 1999, the cost for a three minute telephone call had fallen dramatically.[25] The telecommunications transformation formed the foundation for the expansion of the Internet in Japan.

The Japanese government was forced to abandon its resistance to the Internet and even began to slowly promote those aspects it deemed important. In 1997, the Ministry of International Trade and Industry (MITI)[26] allocated 30 billion yen to electronic commerce pilot projects, attempting to jump-start the somewhat recalcitrant Japanese retail sector.[27] However, until 2000, the government did not really make a concerted effort to encourage the IT revolution nor did it do the one thing it really needed to do to increase Internet usage – reduce the high cost of accessing the Internet by forcing NTT to lower its rates for local calls. (In Japan, telephone users are charged for each local call. They do not, as in North America, pay a single monthly fee that is not tied to usage.) Internet usage in Japan grew but it grew slowly. Ironically, it was the entry of NTT's DoCoMo and i-mode (Internet accessible mobile phones) that caused Internet (albeit a somewhat limited Internet) usage rates to soar. (DoCoMo and i-mode will be discussed in Chapter 3.)

Over the final decade of the 20th century, the government of Japan foundered through a series of economic and political crises. The faltering fortunes of LDP-led administrations (12 Prime Ministers in 15 years) left the country substantially rudderless in times of profound economic and technological change. Bitter infighting within the LDP and between the LDP and its coalition partners resulted in a series of caretaker Prime Ministers and considerable uncertainty at the political centre. Public criticism of the former stalwart departments of the Ministry of International Trade and Industry and the Ministry of Finance shook the foundations of Japan's approach to economic development (which was based on the union of government and industry), and raised open questions about the legitimacy and capabilities of the much vaunted national civil service. That NTT, the state-owned telecommunication system, and the Ministry of Posts and Telecommunications, were widely seen as standing in the way of the development of the Internet, e-commerce and the Information Revolution generally, undercut any suggestion that government might lead the nation through the profound and potentially troubling electronic transformation.[28]

Year after year, government agencies were openly assailed especially by foreign high technology firms attempting to enter Japan for holding back the forces of 'progress'. Japan's Internet usage rates lagged far behind other nations, and critics correctly and enthusiastically pointed out that government policies prevented the country from responding to opportunities. On a wide variety of issues, Internet rates, telecommunications regulations, rules governing foreign companies, and the like, the Japanese government seemed oblivious to the needs and opportunities of the Internet and technological revolutions. The private sector did what it could, working around the myriad barriers imposed by telecommunications policy to create a vibrant if expensive Internet service. But most observers argued that official regulations relegated Japan to second or third tier status in the technology sphere. Early steps such as the call in 1994 for greater efficiency in government using information technology, major discussions of the development of e-government in 1997, and the designation of e-government as a 'millennium project' by Prime Minister Keizo Obuchi in 1999 faltered and brought about little change. Not until 2000, and then only when prodded by a steady decline in the public opinion polls and irrefutable evidence of the gathering importance of information technology, did the government begin to move. The IT initiatives came, in large measure, as part of a general government restructuring, which in turn originated in several consecutive years of economic difficulties. Trade liberalization, a loosening of government regulations generally, financial and regulatory reform, privatization of state-owned enterprises, and associated administrative changes were offered as solutions to the country's lingering economic malaise. These new policies made it easier for foreign firms to enter Japan's marketplace, liberated Internet delivery systems (and sparked a sharp decline in connection prices), and ushered in major changes in the approach to government in Japan. In the midst of this political and administrative transformation, the directionless government of Prime Minister Mori looked for new initiatives which might spark further economic growth and address the country's political and administrative challenges.

   The Mori government seized on IT, albeit with little finesse and even less understanding. In the fall of 2000, the Prime Minister rose in the Diet to deliver a speech which he clearly believed would be the cornerstone of his tenure in office. He spoke in glowing terms of the Information Technology revolution, and promised his countryfolk that Japan would be put on track to become a world-leader in the use of the Internet within five years. Mori offered a vision of major government investments in infrastructure and education, and of a further demonstration of the power of Japanese government–business cooperation, all with a view to ensuring that Japan emerged at the forefront in the digital race. He promised an e-Japan, the equal or superior of the United States, with 40 per cent of the Japanese population online by the end of 2001. 'All people "need to grab the historic opportunity of the IT revolution", Mori said' as he emphasized the government's goal of making Japan the world's most advanced IT power within five years.[29] As Mori observed:

   The 'Japanese IT society' toward which we should aim is a society in which all people can share information and knowledge on the basis of digital information, and freely exchange that information. As such, providing the basic foundation for our society will be high-speed Internet connectivity, through which enormous volumes of digital information 'not only text, but also sound, image and even economic information' can be exchanged rapidly and cheaply.[30]

Mori's stirring speech was rendered less dramatic when it quickly dawned on the Japanese that the Prime Minister really did not know that about which he was speaking. Questioned after his address by reporters, Mori stumbled and rambled, clearly uncomfortable with the inquiries and not at all confident with the technological jargon of the Internet and IT. Pressed to describe his experiences on the Internet, the Prime Minister was forced into the embarrassing admission that he had never been on-line and did not really understand the technology and applications of the Information age. In the following weeks, Mori's political handlers attempted to rehabilitate his reputation by organizing some web-experiences and, in a rather unimpressive attempt to show how open he was to new technologies, had the Prime Minister spend time in Akihabara, Tokyo's electronics district.

Prime Minister Mori was not Al Gore, the American Vice-President who played a crucial role in promoting the Internet. The politically vulnerable and gaffe-prone politician was a far-from-perfect spokesperson for a technological revolution. As his party struggled to hold onto office, however, the Mori administration assembled a far-reaching and dramatic program designed to modernize and digitize government and society in Japan. Unlike his predecessors, earlier Prime Ministers who seemed to have a firm grip on economic and political changes, Mori appeared lost at sea in the high-technology world of the Internet, i-mode, band-width, and silicon chips. In one of the better examples in recent years of the capacity of the civil services to direct government policy, however, mandarins managed to convince Mori and his colleagues of the fundamental urgency behind the Internet revolution. The result, as will be shown, was a series of dramatic commitments to turn Japan into a test-bed for e-government.

Japan-watchers had reasons for scepticism. In the years after the collapse of the bubble economy, when Japan quickly morphed, in the eyes of western observers, from a raging economic tiger to a directionless fiscal basket case, the country found itself with a series of unimpressive and unimaginative political leaders. The once-vaunted Liberal Democractic Party foundered from scandal to scandal, from mistake to calamity, and seemed unable to provide the country with economic direction. Western observers, always quick with advice for Japan, touted the benefits of American-like economic reforms, and demonstrated a level of self-satisfaction and economic confidence which soared during the Internet-driven stock boom of the 1990s. Many observers pointed to Japan's weak start on the Internet and slow adoption of the dot-com revolution as a sign of its inbred conservativism and complacency. Analysts within Japan and overseas urged the country to adopt major structural reforms and, in particular, to seize the economic logic of the digital revolution.[31] Among these watchers were many who saw, in the country's belated enthusiasm for the Internet and digital commerce, a chance for Japan to pull itself out of the doldrums and reinvigorate its domestic and international economy.[32]

While much of this sentiment was America-envy, Japan stood comparatively moribund during the profitable chaos of the early years of the dot.com revolution. There was also a sophisticated sense that Japan's limited investment in advanced technologies for public use threatened to hold the country back. While the rest of the world seemed poised to tackle the 'new economy', Japan seemed wedded to a 1960s vision of government-fed economic growth through funding for public works, low taxes and subsidies for national businesses.[33] By the mid-1990s, however, change had started. Private Internet companies entered the marketplace, NTT lost its hammerlock on the domestic telephone market, and international firms (using call back services) provided price competitive alternatives to the shockingly high Japanese domestic and long distance telephone rates.[34] The dot.com collapse that occurred in late 2000/2001 in the United States also took away some of the western hubris and forced both foreigners and Japanese to realize that there were some positive aspects to moving a little more slowly with this new revolution.

The July 2000 G-8 Summit held in Okinawa focused on IT issues, and provided the Prime Minister with a high profile entree into the field. On the heels of the summit, Mori placed Chief Cabinet Secretary Hidenao Nakagawa in charge of the development of an IT plan for the government and urged him in conjunction with other cabinet members (the IT Strategy Headquarters group) to develop rapidly a national strategy.[35] To further this end, the Mori administration also established at the same time an Information Technology Strategy Council (ITSC) to develop a long-term plan for the country. To form the ITSC, Mori assembled a high powered advisory group of 20 leading IT experts, primarily from the private sector (headed by Sony Chair and CEO Nobuyuki Idei and including Mari Matsunaga, who led NTT DoCoMo's i-mode project, Nayouki Akikusa, president of Fujitsu, KDDI president Yusai Okuyama, NTT President Junichiro Miyazu, Bank of

Tokyo Chair Saforu Kishi, and Professor Heizo Takenaka of Keio University, among others) and gave them a mandate of helping the government devise a formal IT strategy to be completed by November 2000.[36] Japan, it seemed, was prepared to work at 'Internet speed'.

Government followed the work of the Council assiduously, with cabinet ministers and the prime minister in regular attendance at their meetings. The ITSC put foward a full range of initiatives, from subsidies for Japanese firms to major foreign aid expenditures in the area. It proposed a goal of providing 30 million households with 'constant access to high-speed networks and 10 million households access to super high-speed networks within five years'.[37] In addition, the ITSC proposed that within one year all Japanese people should have easy low cost access to the Internet.

The government did more than just listen to the ITSC. Mori's administration committed 1 trillion yen to IT initiatives, promising such diverse measures as the establishment of a computer network connecting 4,000 schools, Internet training for seven million adult Japanese, regulatory changes to encourage greater use of IT in business and promote online government services, financial assistance to start IT-related businesses, the introduction of a high speed phone network to smaller Japanese cities where fibre optic lines were not available and research and development monies to explore the potential of IP Version 6 which allows the Internet to be accessed from cars and electrical appliances.[38] The government promised to move government services on line, bring computers into general use in its administrative offices, deregulate Internet fees, and support increased competition in IT. Japanese authorities had a long way to go. Offices were poorly equipped, staff had little computer training, and Japanese office practices were not easily adapted to the e-government environment.[39]

At the end of November 2000, the IT Strategy Council submitted a report on a Basic IT Strategy for Japan. As the report outlined:

> On the threshold of the 21st century, Japan must take revolutionary yet realistic actions promptly, without being bound by existing systems, practices and interests, in order to create a 'knowledge-emergent society,' where everyone can actively utilize information technology (IT) and fully enjoy its benefits. We will strive to establish an environment where the private sector, based on market forces, can exert its full potential and make Japan the world's most advanced IT nation within five years by: 1) building an ultra high-

speed Internet network and providing constant Internet access at the earliest date possible, 2) establishing rules on electronic commerce, 3) realizing an electronic government and 4) nurturing high-quality human resources for the new era.[40]

The report highlighted four areas for immediate government attention: network infrastructure and Internet competition policies, improved conditions for e-commerce, a national commitment to e-government, and the development of the human capital necessary to create and exploit the IT systems.

The government showed every sign of being determined to act promptly. In October 2000, the government introduced sweeping legislation – revising 50 different laws – to facilitate the expansion of e-commerce. The laws attacked the fundamental premises of business to consumer transactions, freeing companies to rely on digital means of purchasing and payment.[41] Not all the measures were well-received. The Mori government remained wedded to direct personal subsidies for IT and proposed the establishment of a personal IT voucher. Vouchers distributed to some 30 million adults were to be used for basic computer literacy training sessions, and sought to improve the computer knowledge base in the country. In the tangled web of Japanese coalition politics, however, the plan failed to secure support within the ruling parties, and the initiative was withdrawn.[42]

In November 2000, the Lower House of the Diet (Japan's parliamentary body) passed the Basic Law on the Formation of an Advanced Information and Telecommunications Network Society (IT Basic Law), based on the IT Basic Strategy report. This new law sought to create a broadband network infrastructure, promote e-commerce by eliminating or easing some of the estimated 733 regulations and 124 laws currently hindering the conduct of business on the Internet and to protect the privacy of on-line information.[43] The IT Basic Law required government to develop specific goals relating to the desired IT revolution in Japan and to set deadlines for achieving those goals. It encouraged all levels of government to provide as many services as possible on-line and to train the population in IT technologies. The bill also outlined provisions for the establishment of a new task force with the responsibility for making Japan an IT capable country.[44] The law passed the Upper House in December and went into effect in January 2001. This approach, interestingly, mirrored efforts in the 1950s and 1960s to stimulate development in emerging sectors (such as the earlier Basic

Law for the promotion of the electronics industry) through government signalling and investment.[45]

It is crucial to note the relative absence of organized protest, in the Diet or outside the legislature, to the Mori plan. Whatever the public and political opposition thought of the Prime Minister, and public opinion polls documented his staggering unpopularity, there was widespread agreement that Information Technology was a crucial element in determining Japan's economic future. In fact, at a time when virtually every step taken by the Prime Minister and his cabinet was criticized and assailed by political opponents and the public at large, the IT initiative moved through the Diet with relative ease and found surprising public support. Not everyone was completely confident that the plan would work and many worried that politics would interfere with the implementation of the IT plan. As one *Japan Times* columnist observed:

> Creating infrastructure is the right idea, but I have little faith in this government's ability to leave it at that. My fears were sharpened by the revelation that under the new plan, sewer systems, a time-tested serving of pork to construction companies, count as IT-related because cables can be placed in them.[46]

In early January 2001, the government established a new information technology strategy council (as per the Basic IT bill), referred to as the IT Strategy Headquarters (ITSH) and consisting of 18 cabinet members and ten representatives of the private sector, local government and the academy. According to Mori, the ITSH would be at the heart of Japan's attempts to become the world's most advanced IT nation. It would be responsible for pushing forward high technology related legislation and promoting the objectives outlined in the IT Basic Law. At their first meeting, the ITSH agreed that the e-Japan strategy should promote IT in the priority areas which had been highlighted by the ITSC (high speed network infrastructure, electronic commerce, electronic governments and human resources).[47] A task force to examine information security was also set up under the council. On 22 January 2001, the ITSH announced the e-Japan strategy which set objectives of realizing e-government by 2003 and achieving an e-commerce market size of over 70 trillion yen along with a commitment to promoting education and learning about IT and, most importantly, forming the world's most advanced IT network. (Included in the e-Japan program were tax incentives for the development of high-speed fibre optic, cable and mobile Internet systems.) On 29 March 2001, the e-Japan Priority Policy Programme was established to put in place the steps needed to achieve the e-Japan objectives by 2006.[48] For fiscal year 2001 alone, the government committed 2 trillion yen to implementing the programme.

The e-Japan programme set a number of goals to be achieved by 2003, most of which were reached. The most important of these were increasing Internet access and reforming regulatory frameworks to increase competition in Japan's telecommunications industry. That the virtual monopoly on local calls held by NTT was more than partially responsible for the low levels of Internet use in Japan was well known. By late 2000, the Ministry of Posts and Telecommunications, Japanese and foreign businesses and even many members of the ruling Liberal Democratic Party were pushing for reforms to NTT.[49] The government began developing plans to completely break the telephone company's stranglehold on domestic telephone service and to permit greater competition in this vital sector.[50] Earlier in the year, NTT had been broken up into NTT Holdings (which included NTT DoCoMo), NTT Communications (long distance calls) and NTT East and NTT West (local operators) but calls for further changes, particularly lower local call rates, continued. In November 2000, NTT East and NTT West announced a slight decrease in local call rates – from 10 yen (approximately 7 American cents) to 9 for a three-minute call.[51] Flat rate Internet connection fees (set at about 5,000 yen a month) became available in December 2000. Competition in the local call (and therefore Internet connection) arena increased. KDDI (an amalgamation of three phone companies) and NTT Communications discussed entering the market.[52] As new developments appeared on the scene, like NTT's L-mode home Internet service (providing Internet via telephone lines and using a telephone device instead of a personal computer), the government threw its weight behind the initiatives. Toranosuke Katayama of the Ministry of Public Management, Home Affairs, Posts and Telecommunications, showing the new government's determination to expedite telecommunications innovations, pressed the companies to work through the regulatory difficulties in order to bring the service into operation as quickly as possible.[53] The government established high profile targets, aimed largely at ensuring the public greater access to government services and information on-line (Table 2.1). In the process, they left themselves with the formidable challenge of achieving the highly touted goals.

Table 2.1   E-Japan targets[54]

On-line applications for selected government services	End of Fiscal 2002
Use IT installations to modernize cargo clearance systems	Fiscal 2003
Integrated data-base of government procurement	Fiscal 2001
Internet bidding for government procurement	Fiscal 2003
Public works projects for on-line bidding	Fiscal 2001–2004
On-line tax payment	Fiscal 2003
IT applications of administrative functions (selected)	Fiscal 2002
Government network (central–local)	Fiscal 2003

The government followed through on its promise to reform business operations to suit the technology of the Internet. Japan's e-Signature Law went into effect in April 2001. This law aimed to have a marked impact on e-commerce in Japan. Prior to the implementation of the law, business had to be conducted with written documents and face to face sales, making business over the Internet virtually impossible. The new law allowed for electronic signatures and electronic authentication to be as legally valid as written signatures and seal impressions.[55]

As a central part of the Mori plan, the government committed itself to reconceptualizing government in Japan. The administration faced a formidable challenge. Japan's civil service was, compared to other leading industrial nations, poorly served in terms of computers. As of March 1999, there was one personal computer per 1.6 employees at ministries and agencies and in national government agencies overall, there was one computer per two employees.[56] Most government services remained, at century's end, locked in the inefficient and time-consuming paper-based systems of the past. Japan had one of the first world's lowest rates of computer use among government officials, and very few attempts had been made to capitalize on the advances made in the area of IT. Government agencies provided web-sites (many key ones translated into English), but virtually no web-enabled services beyond basic electronic pamphlets. Email was not in widespread use within administrative circles, and e-commerce business models had not yet been applied to government operations, as was already happening in many western nations. The Keidanren (Federation of Economic Organizations) estimated in May 2001 that of the 265 million monetary transactions made by the government annually, 90 per cent of revenue transactions and 30 per cent of expenditures were still handled on paper. Moving these transactions on line would, the organization claimed, save consumers, financial institutions and government agencies a combined total 100 billion yen.[57]

With a boldness not otherwise associated with his administration, Prime Minister Mori announced that this would soon change. Within five years, he promised, government operations would be fully computerized, a potential boon to the country's manufacturing sector. Furthermore, the government committed itself to investigating e-government initiatives, ranging from the payment of taxes to information services. These activities would, it was clear, cost a great deal of money, require extensive training, and involve the government in a major transformation of its civil service. For this sector, much more than private business, had remained isolated from the revolutionizing changes associated with IT and appeared wedded to traditional service models. If the government succeeds in adopting digital solutions for government, it will go a very long way toward inculcating a technological culture within the country.

The vision of e-government goes well beyond the transfer of administrative functions onto computers and the Internet, important though this may be. The Mori plan foreshadowed numerous initiatives in other areas of government responsibility. Education would, the government declared, have to be altered to both provide the skilled technicians required for the information age and to capitalize on the educative possibilities of new media (CD-ROM, the Internet, and computer-based educational programs). Tele-health, the provision of technologically mediated health care solutions, held enormous promise, particularly for the country's growing population of seniors. Major investments in this field had the potential to improve home-based health care delivery and to thus reduce pressure on the more expensive hospital system.(The Internet has also created a substantial group of cyberchondriacs, who use the world wide web to gather information on illnesses, cures and treatments.[58]) Furthermore, telehealth could, if successful, permit advanced service delivery into rural and remote regions,[59] and greater connectivity between specialist and general practitioners,[60] thereby addressing a significant shortcoming in the current health care system. One of the first significant moves in this area came with the establishment of CareNet Inc, an Internet health care business to business (or hospital to hospital) application. Faced with an ageing population that is stretching the country's health care resources, CareNet allows hospitals to allocate patients and resources between different publicly-funded facilities, thus enabling more efficient use of limited facilities and funds.[61] The government has moved on a variety of fronts,[62] gradually bringing the country more closely in line with other industrialized nations.[63]

Discussions about the potential evolution of digital government ranged widely. In January 2001, the Ministry of Public Management, Home Affairs, Posts and Telecommunications announced that they had started work on legislation to permit electronic voting; the legislation was passed early in 2002. The plans called for the testing of electronic voting at the local level, with an expansion to national politics if the early experience warranted.[64] Similar initiatives were already underway in the United States and other countries. In this area, as in many other fields of e-government, Japan was continuing to play catch-up.

The first significant rollout of e-voting occurred in Niimi, Okayama Prefecture, in June 2001. The local government used touch-screen technology, providing classes and instructions in advance to ensure that voters knew how to use the machines. This is not e-voting at a distance, long touted as the eventual use of the technology. Instead, electors had to visit a voting station. They were given a card (carrying individualized codes) which they used to activate the voting screen. A special pen was then used to vote on the touch screen. Niimi discovered that the system was much faster than a conventional paper ballot, worked efficiently, used many fewer municipal employees and cost a great deal less money. The set-up was costly, a fiscal challenge which deterred other municipalities from copying the system, but it seemed efficient, secure and easy to use.[65] Not all e-government implementations have been as costly and comprehensive. Officials in a small rural community in northeastern Japan. In this instance, the limited resources of government encouraged the authorities to capitalize on the Internet as a means of incorporating 'on-line citizens' into the work and political life of the local district.[66]

From the national government's perspective, the Niimi experiment represented only the first phase of a long-term shift toward electronic voting. In the first phase, priorities rested on municipal elections and required voters to come to the traditional electoral stations. If the experiment expands, the government foresees having a greatly expanded number of voting booths, including voting machines at train stations and other high traffic locales. As technology and public will permits, the government intends to eventually extend the system to include at-home voting, although there is considerable scepticism about the likelihood of this step arriving in the near future.

Surveys conducted by the Japan Research Institute in 2000 and 2002 reveal the growing commitment to, and understanding of, the use of information technology in government. The investigation revealed a rapid sophistication of e-government in Japan. In October 2000, only

28 per cent of government offices surveyed had all of their computers hooked up to a network. By March 2002, that percentage had jumped to almost 62 per cent. Similarly, in 2000, only a quarter of government offices reported that all employees had a personal email account, which could be used for internal and external communications. Sixteen months later, over half (52.9 per cent) had this capability. In 2000, of the agencies surveyed, almost 64 per cent had only one computer per division. By 2002, that number had fallen to less than 40 per cent (39.3 per cent). Not surprisingly, the officials also noted that they lacked the budget to implement their full IT strategies and had difficulty finding funds to pay for maintenance and operational costs. It was also clear that managerial understanding had improved significantly and that major inroads had been made into meeting IT needs (see Table 2.2). The JRI survey also provided revealing insights into government plans for expanding e-services, showing a pronounced interest in electronic purchasing, smart-cards and the provision of local information over the Internet (Table 2.3).

In the last years of the 20th century, Japan lagged far behind other leading industrial nations in the development of e-government. A study of e-government in 22 countries conducted by Accenture in January 2001 put Japan in the lowest category, 'platform builders'- countries with low levels of online service – along with Brazil, Malaysia, South Africa, Mexico and Italy.[67] The slow adoption of the Internet in homes put a brake on public demand for the provision of e-based services and let the government off the hook. Significantly, the development of mobile Internet use (via the *keitai*) involved virtually no government presence and no significant government support. This was an initiative of NTT-Docomo and its competitors and did not rep-

Table 2.2   IT difficulties facing local government
(Survey of local government officials, 2000 and 2002,
% responding, multiple answers permitted)

Difficulty	2000	2002
Securing budgets	77.7	71.4
Organizational structure	63.8	45.5
Infrastructure	45.7	19.5
Knowledge gap	33.0	36.4
Needs gap	33.0	19.5
Skills gap	48.9	42.9
Managerial understanding	36.2	5.2

Source: Japan Research Institute Survey, reported in Japan Inc., Aug. 2002 (p. 45).

Table 2.3   **Preferred e-government services for development
(survey of local government officials, 2000 and 2002,
% responding, multiple answers permitted)**

Service	2000	2002
Electronic procurement	50.0	56.5
Electronic public auctions	50.0	63.0
Local smartcard services	75.5	57.6
Community services	38.3	41.3
Local web-sites	71.3	67.4
Senior community services	26.6	22.8
Local e-learning	41.5	33.7

Source: Japan Research Institute Survey, reported in Japan Inc., Aug. 2002 (p,. 45).

resent a centrally stimulated effort to make the Internet more generally acceptable. The Mori government's sudden reversal – its discovery of the potential and importance of Information Technology – was viewed with some scepticism but, even more, also with a sense of relief. The Japanese could see what other nations, principally the United States, had done with the IT revolution and were anxious to see the new technologies at work in their country. The 2000 Mori initiative had the potential to accelerate Japan's race to get back into the game. Government, having largely ignored the most important technological innovations in the past two decades, prepared itself to lead the nation into the forefront in the digital revolution. Many hoped the government's latter-day conversion to IT would be sufficient; most Japanese observers were uncertain as to whether it was soon enough, large enough and comprehensive enough to make a real difference.

At much the same time that the Mori government recognized the importance of the Information Revolution for Japan, it also concluded that the initiative held enormous importance in the international field. They were pushed to this realization, in large measure, by Japanese technology firms, which were anxious to establish a greater presence in the global and, in particular, the Asian technology markets. North American firms had a formidable lead. Companies like Oracle, Intel, Cisco, Nortel, Sun Microsystems and others presented themselves as the logical technology providers for the emerging economies in Asia. Furthermore, American and English-language web-sites and e-business had blazed the trail on the use of the Internet for communication, e-commerce, and technological innovation. Japan had, in the late 1990s, a thin and unimpressive presence in many of Asia's technology markets.

Japan's technology industry, however, was internationally impressive and determined to break into the fast growing Asian markets. Most Asian countries (except for Singapore and, to a lesser extent, Hong Kong, South Korea and Taiwan) had initially responded slowly to the promise of the Internet and the Information Technology revolution. National governments, reeling from the Asian financial crisis after 1997, lacked the national will and fiscal resources to mount major infrastructure initiatives. And while the World Bank, United Nations and other organizations worried openly about the growing digital divide between industrial and developing nations, there were few resources in the region available to respond creatively to the need and opportunity. There were significant undertakings, particularly in South Korea, China, Taiwan and Hong Kong, in response to the development of the Internet, and the region quickly produced its share of dot.com promoters (and overnight millionaires) and technology leaders.

This environment fitted nicely with Japan's strategy for reasserting itself as a technological force in the Asian region – a long-standing national aspiration that had found renewed life in the face of Japan's strong and determined regional fiscal response to the Asian financial crisis. The Mori government was determined, through 2000/2001, to ensure that technology followed the yen into the developing countries in Asia. When Prime Minister Mori met with his counterparts from the Association of Southeast Asian Nations (ASEAN) for the annual Japan-ASEAN summit in Singapore in November 2001, he committed the majority of a $15 billion (US) information technology aid allocation over five years to the development of IT projects in Asia. This amount came out of a total annual Foreign Aid budget of approximately $15 billion a year, a budget under severe downward pressure in government circles. As part of this $15 billion IT allocation, Japan also pledged to establish thirty IT centres using satellite and other technologies to provide long distance training programs for developing countries. The foreign aid represented a substantial investment in regional infrastructure development. It also provided a strong indication of the Japanese government's determination to ensure that national information technology firms played a significant regional role. The two-pronged goal of this initiative, after all, was to ensure that the developing companies of Asia participated in the IT revolution and that, to the greatest degree possible, they did so with Japanese technology.

Japan's government is serious about Japanese technology, Japanese companies, and Japanese services serving as the foundation for the Asian IT revolution. Largely removed from western view – although the

initiatives are scarcely secret – the government continues to push initiatives designed to integrate national technology and companies into
the very fabric of the Asian Internet expansion. They have done so
through development aid, by encouraging trade, and by working to liberalize regulatory regimes between countries. Early in 2001, Japan and
Singapore were negotiating a free trade agreement designed to facilitate
greater cooperation in IT and e-commerce and to make trade in digital
products much easier.[68] Such initiatives are founded on the simple
belief that IT holds the key to long-term international competitiveness
and that access to the large and growing Asian market is central to
Japan's economic viability.

In April 2001, Mori's political unpopularity resulted in his replace-
ment by Junichiro Koizumi as Japan's prime minister. Koizumi, hailed
by the public as the reformer Japan needed, pledged to continue the
governmental IT initiatives. In his maiden speech to the Diet, he said
'you are all aware that we have set ourselves the ambitious goal of
making our nation the most advanced IT state in the world within five
years. In order to ensure we achieve this goal, I will steadily implement
the 'e-Japan Priority Policy Program'. As a mid-term goal, I also intend
to formulate the 'IT 2002 Program'.[69] In this area, unlike most other
government responsibilities, Koizumi did not need to jump-start the
LDP administration. Koizumi reiterated his commitment in his speech
to the Diet in September 2001 but provided few specifics. 'In relation
to information technology, we have formulated the e-Japan 2002
Programme as an interim target and have accelerated the move toward
becoming the most advanced IT nation in the world. I will work inten-
sively toward achieving such measures as e-government that enables
applications and notifications to be made from homes and offices'. [70]

To emphasize his familiarity with the Internet and reach out to the
public, Prime Minister Koizumi launched a weekly e-mail magazine.
The electronic magazine, titled Lion Heart (in reference to the Prime
Minister's 'tousled mane of wavy hair and his promise of bold reform'),
proved to be immediately popular.[71] By the time the second issue came
out, Lion Heart had 1.8 million subscribers and by January 2002, about
2.3 million people received the weekly e-magazine.[72] The email magazine and the Internet generally fit perfectly with Koizumi's image as a
radical, new-age thinker, in tune with younger people and with the
economic and social challenges facing the country. This simple initiative provided a strong message that Japan's administration might actually tackle the e-government challenge. In general, the political parties
in Japan moved quite slowly to capitalize on the Internet, although

there has been increased emphasis placed on the new technologies after 2001.[73]

As other countries, South Korea being the best example, have demonstrated, governments can spark a nation-wide migration towards internet utilization by moving their administrative activities on line. Japan moved slowly in this regard, continuing to rely on expensive, labour intensive systems. The government began moving toward greater administrative use of the Internet in 2002/2003, implementing a series of 'catch-up' initiatives to permit the completion of selected tasks, such as tax filing, requesting documents, and otherwise interacting with government offices on line. The decision in 2002 to out-source much of the work to the private sector brought in considerable foreign investment and convinced many Japanese firms to make significant commitments to this once marginal field. One measure of the seriousness with which the business community viewed the government's initiatives was the degree to which major companies, typically through complex alliances, stepped forward to secure contracts related to e-government initiatives. The largest e-government firms in the country (NEC, Fujitsu, Hitachi, NTT Data) dominated the field. A number of other technology firms, drawn together by Sun Microsystems and Fuji Xerox, created the Open Source Technology Consortium in early 2003 to compete for what was viewed as a lucrative and growing market for e-government solutions.

The task of establishing the country as a world-leading IT nation within five years echoed 1960s promises of rapid economic growth and income doubling targets. The Basic IT Strategy document set a goal of reaching 30 million households with high-speed service (30–100 mbps). The Council backed the further development of Internet access through digital appliances (other than PCs) and the expanded use of mobile Internet. The Council urged, and the government acknowledged, that the country needed to recruit over 30,000 foreign IT engineers and researchers combined with a renewed effort to develop home-grown talent[74]. This was, indeed, a bold vision, particularly for a country mired in a long-term economic downturn.[75]

It needs to be said, after all of this, that Japan is far from alone in believing that digital delivery and support systems could revolutionize public administration. Singapore is already perhaps the most wired government in the world, and countries as far apart as Finland and New Zealand, Hong Kong and Canada, France and Sweden, have made substantial commitments to applying information technologies to the provision of government services. While in most cases, this relates largely to the availability of government information and download-able official application forms, some governments have taken more dramatic steps. In the latter cases, citizens can pay taxes, fees and fines on-line, make formal application for jobs and government services, contact officials, and register their thoughts on government initiatives. A few jurisdictions are contemplating, as is Japan, e-voting, electronic town-hall meetings, and sophisticated and widely-accessible land registry and public information services. It remains to be seen if e-government is a public administration version of the dot-com revolu-tion – heavily promoted, oversold, and unable to deliver on the appar-ent promise of the new technology – or an illustration of the world's continuing ability to under-estimate the full impact of the economic, social and political dimensions of the Information Technology age. The cost-saving possibilities of this sector are considerable, and encour-age government officials to marshal support for initiatives in this area. Early assessments suggest that the potential for e-government is much like that of business to business e-commerce. The flashy government to citizen services are less likely to flourish than internal e-government implementations. The latter have the potential to revolutionize the way in which government conducts its affairs.

Japan has a critical advantage over many other countries in this field. Should the government proceed with a significant implementation of e-government, the long-term tolerance of top-down initiatives in Japan provides a potentially vital foundation for the successful transfer of government programs onto digital platforms. Highly centralized states like Japan, Singapore, and a handful of other industrialized nations have opportunities that do not exist in more diffuse, democratic and citizen-oriented nations. The Japanese government has the option, in such fields as education,[76] health care, and general government ser-vices, of dictating to its employees and clients the format for govern-ment program delivery. At present, the Japanese tolerate an efficient, employee-heavy structure in many areas, and considerable rigidity in the provision of educational and health care services. Following the Singapore experience, it is likely that the Japanese would accept – more readily than Americans, Canadians, Australians or the British – a shift to e-based government programmes.

The country's politicians, of all political stripes, have repeatedly indicated their determination to push Japan to the forefront of the IT revolution. If they do so, the engine of this transformation will likely be e-government, with Japan's central administration providing a strong impetus to the digitizing of the nation-state. Japan has the technological capability. It has clearly indicated its willingness to spend large sums of money both on IT and government-driven social and economic change. It sees IT as the cornerstone of the new economy, and has declared its intention to modernize and streamline government service delivery through the use of technology. Although political uncertainties can undermine government resolve, the potential exists for e-government to lead Japanese society into the next phase of the IT revolution.

Governments, in Japan and elsewhere, have two critical roles to play in the roll-out of Information Technology and the adaptation to the Internet. In the first, they have been challenged to enable the business sector responds to the opportunities in the rapidly unfolding field and, particularly through the restructuring of telecommunications services and policies, to ensure that companies and citizens have affordable, reliable access to the Internet. For the second, governments face the challenge of harnessing the potential of the Internet to official services and operations. As with e-commerce, significant benefits will accrue to the governments and countries which manage both of these assignments ably and quickly. Japan is not a leader in either of these areas, and has a great deal of catching up to do to convert the country into an Internet-ready nation.

The Mori and, later Koizumi, plans for Japan's rapid transition into an Internet powerhouse – political rhetoric is awash with references to becoming the leading Internet nation by 2005[77] – would be more impressive if they did not replicate statements by governments around the world. In country after country, particularly in the industrial world, governments have declared their determination to provide a suitable platform for an Internet-enabled society and their commitment to taking government services on-line. The most successful nations in this regard – the United States, Singapore, the United Kingdom, Canada, Finland and Sweden – have a formidable lead over Japan. This means, of course, that Japan stands to learn and benefit from the successes and failures of these other nations. It also means that high-sounding claims to Internet supremacy by Japan remain unfilled promises. It remains to be seen if Japan's government has the will, resources and commitment necessary to ensure that the country makes the Internet transition and leap-frogs its international rivals.

3

The Keitai Revolution: Mobile Commerce in Japan

Throughout the late 1990s, much of the industrial world salivated at the prospect of electronic commerce, the marriage of technology and business that threatened to undo the verities of retail, wholesale and financial operations around the world. It is easy, even a few years on, with the excitement dissipated by the step-wise failure of many dot com visions, to forget the excitement that reigned through the last half of the decade. E-retailers like Amazon.com threatened to undermine the entire book-selling sector. On-line financial services, from bill payment to stock trading, transformed the banking and stock broker-age business. Auction sites, like eBay, reintroduced barter into the western economy. Entertainment companies, newspapers, and maga-zines rushed on-line, determined to find market share and hefty returns among the *digiteratia*. Japan lagged well behind in this com-mercial explosion, to the point that the country was ridiculed by those who 'knew' where the digital revolution was heading.[1] That the nation was not sophisticated enough to use credit and debit cards and thus participate in on-line commerce simply indicated how far behind Japan had fallen in the digital race.[2]

Critics of Japan's progress, however, forgot one basic truth to the Internet: the system hates barriers. The core concept of the world-wide web is that information flow is virtually continuous. If one access route is blocked or congested, the data simply finds an alternate route. For the Japanese, the barrier was corporate, not technological: Nippon Telegraph and Telephone's (NTT) complex regulations governing inter-nal telecommunications charges.[3] With access charges running five or more times higher than most industrial nations, home computer Internet use was prohibitively expensive. When North Americans were mulling over 'all you can surf' packages, and playing with the prospect of free computers and Internet access service for those willing to be bombarded by digital advertising, Japan limped along with slow connection speeds, expensive charges, and comparatively low rates of usage.[4] Personal e-commerce, as a consequence, had few chances of early success.

Clearly, the personal computer-based Internet was ill-suited to Japan. Aside from the charges involved, Japanese homes tend to be very small, particularly in major urban centres, and there is little dedicated space available for a full-size computer. Similarly, commuting and work schedules are notoriously intense in Japan. Most office workers endure lengthy commutes, leaving home early in the morning and arriving back late at night. Children fare little better, for their regular school hours are compounded by the wide-spread practice of attending *juku* (after hours) schools in the evenings and on weekends to improve their test scores and increase their chances of getting into a better university. Compared to the leisurely lifestyles of the North American, European and Australasian web-surfers, the average Japanese family had little time to spend on a home computer.

Until 1999, western observers added these factors together and concluded that Japan was doomed to remain far behind the rest of the world in the adoption of the Internet and the related development of e-commerce. Imagine the surprise, then, when the turn of the century witnessed the advent of one of the most dramatic technological adaptations the world has seen. (Enthusiasm for digital music swapping software like Napster among North American teenagers paled in comparison to the speed of the Japanese response to their technological opportunity) The development that turned the country's Internet prospects on their ear was the antithesis of the heavy, site-bound, powerful North American home computer. What sparked the Japanese revolution was the *keitai*, a small handheld portable telephone, the foundation of the country's leap into world leadership in mobile commerce.[5]

Mobile phones have become increasingly commonplace around the world, starting generally with business users and, as usage charges and parental resistance declines, gradually expanding to the general population. While North American usage has expanded slowly, bogged down in debates about standardized protocols, Scandinavians switched to the new telephone systems with enthusiasm. Nokia, the Finish firm, emerged as a world leader in the field and quickly developed a reputation for producing high quality and technologically innovative products. By the mid-1990s, Japan was likewise swept up in the enthusiasm for mobile phones. Standard phone service, operating through the

cumbersome NTT system, developed more slowly than demand warranted. But Japan had a unique advantage – the tightly-packed
30 million residents of the Tokyo district. This area of wealthy, active,
technologically savvy consumers has, since the 1980s, been among the
most innovative business markets in the world. Life moves at a frantic
pace in this sprawling city, as commuters clamber off and on subways
and race through the maze-like streets. Mobile phones were extremely
popular in Tokyo, even when rates were uncommonly expensive, and
through the 1980s, the phone replaced the camera as the main accessory of the stereotypic Japanese. Local firms realized, however, that the
population density created new options. Different communication
systems, based not on the expensive satellite or microwave technology
but rather on line of sight,[6] low power devices, had much more potential here than in the wide spaces of North American cities. For Japanese
firms, the combination of unique local conditions (for those who have
not been to Tokyo, it is important to realize that it is extremely challenging to find one's way around the city except by subway; even taxi
drivers have difficulty) and population density created special commercial opportunities.[7]

The result, by the mid-1990s, was widespread enthusiasm for
personal telephone system (PHS) phones. These tiny phones became
popular immediately. Usage fees were cheap, and companies gave the
phones away to anyone who signed up for a long-term contract.
Adults, teenagers and even young children carried them. At a time
when cellular phones remained largely the preserve of the business
community in much of the world, Japanese consumers embraced
wholeheartedly this less powerful but perfectly suitable local solution.
Telephones and the Internet are not the same, however, and few
outside observers viewed the tiny mobile phones as a challenge to
western domination of the Internet. While North American and
European firms pressed ahead with their efforts to create wireless information services (pagers with downloadable stock prices, global positioning systems for automobiles), several Japanese firms focused their
attention on the unique metropolitan markets in their country.

The introduction of the mobile phone-based Internet[8] in February
1999 overturned the foundations of the Japanese Internet.[9] The commercial revolution started, ironically enough, with an NTT subsidiary,
DoCoMo.[10] (Although a rival telecommunications company, J-Phone,
was the first to bring the mobile-phoned based internet to market, it
was DoCoMo that really capitalized on the technology.) DoCoMo
came to market with a very simple device: a portable telephone with a

small screen (about 11 lines) that allowed you to send e-mail and access a few internet sites. The system was named i-mode with the i standing for information. 'You push the button that says "i", and you change the mode – that's all', explained DoCoMo executive, Mari Matsunaga.[11]

Engineer Kei-ichi Enoki took on responsibility for the development of what was to become i-mode. The service allowed subscribers to do more than just talk on their mobile phones, they could also receive data. Enoki recruited Takeshi Natsuno, previously of Hypernet, one of Japan's first Net startups, and Mari Matsunaga, former *Recruit* (a classified-ad magazine) editor-in-chief. These three people are credited with DoCoMo's success. (Both Natsuno and Matsunaga have written books about their experiences in developing i-mode. Natsuno's *i-mode Strategy*[12] is only available in Japanese but Matsunaga's *The Birth of i-mode* has recently become available in English.)[13]

Matsunaga's particular strength was understanding the technological needs of ordinary people. Matsunaga describes herself as something of a technophobe – her college degree is in French literature and when she arrived at DoCoMo she did not own a cellphone or understand anything about the internet. But, or because of this, she understands what the average person wants from technology.[14] Matsunaga insisted on the importance of content and of ease of use. 'Customers couldn't care less about technology, she argued; they wanted usefulness.'[15]

DoCoMo's i-mode phone, therefore, focused on practicalities. To begin with, the handsets have more readable screens and longer lasting batteries than typical mobile phones. They are small, light and most importantly, unlike personal computers or WAP phones or Palm Pilots, i-mode phones do not require a dial-up connection; they are always on. Secondly, i-mode did not endeavour to reproduce the full Internet on a hand-held phone, but rather offered consumers connections to specialized services. Subscribers initially could only access a limited number of sites, DoCoMo made sure that those sites were of the highest quality. DoCoMo worked with a small number of Japanese firms to prepare basic web-sites, designed to a different protocol and prepared without the heavy traffic graphical content that would otherwise slow the system down. However, according to DoCoMo's president, Keiji Tachikawa, the company realized that once consumers visits a poor quality site, their confidence is damaged and it is very hard to restore. So, right from the beginning the company's focus was on supplying fresh exciting content, sites to which users would wish to return time and time again. (Computer games that must be played many,

many times before you can win is a good example of this kind of 'sticky' site.) The number and array of sites is continually increasing, keeping both current and prospective users interested.

I-mode was first marketed as a new and fun service for the mobile phone – an easy-to-use complement to the voice function – not as the Internet over the phone. This ensured that users had moderate expectations and were therefore not envisioning something beyond what i-mode could deliver initially. Customers were pleased by what they now could do with their phones rather than being disappointed by the sites they were unable to access. A recent global ethnographic study of wireless use backs up this approach. The principal anthropologist of the study, Robbie Blinkoff, states 'We got inside the minds of consumers and what they are saying to wireless manufacturers and marketers is don't over-promise what the devices can do and make it easier for me to use them.'[16] I-mode did just that.

Content and convenience have been key to i-mode's success. As Enoki explained,

> PCs are like department stores. ... 'They have a wide selection of content, including excellent graphic images. If you decide to make a visit, you can have a good look at what's on offer. You can stay as long as you like and explore different sites at your leisure. Mobile phones are more like convenience stores, where only a selection of goods are on display in the limited space available. The contents have to be simple, but the convenience comes from the fact that they can be accessed at any time. We created a mobile distribution platform that gave content providers a basis on which to build a business. That was the starting point of the i-mode business model. What also made the i-mode system so uniquely attractive were the usage costs.[17]

Mobile phones are typically very expensive, with charges based on the connection time. In other words, they mirror the costly and discouraging NTT home Internet system. Although initially there was a transmission charge per minute, DoCoMo quickly replaced this system with a small monthly fee (as early as 2001, it was only 300 yen – less than $US3 – a month) and a new system based on downloading packets of information. Charges are based on the amount of information downloaded, with the costs mounting up in tenths of a yen, rather than deriving from the number of minutes the user was connected to the system.[18]

No one, including DoCoMo properly estimated the consumer demand for the new service. Within months, the number of mobile phone users accessing the Internet in Japan had sky-rocketed. Amazingly, by April 2000, just over a year after the technology had been introduced, DoCoMo had over 5 million users. Rival companies, J-Sky (with its J-phone), KDDI (with its EZweb phone) and DDI Pocket's 'H' (pronounced 'edge') system ranked far behind in subscriptions, but their existence heightened competition and ensured that more web-design companies and e-commerce firms targeted mobile phone web-site development. By mid-2000, when Japan became the first country in the world to have more mobile phones than land line phones and 15 million Japanese had access to the Internet via their mobile phones, the country's technological and communications foundation had been transformed. While government, Internet Service Providers and companies continued to promote the idea of home-based computer use, momentum had clearly shifted to the *keitai*. As of April 2001, 37 million Japanese subscribed to mobile Internet services, 24 million of them through DoCoMo. About 43,000 new subscribers were signing on every day, 1.3 million month.[19] By 2003, estimates indicated that over 45 million Japanese used Internet compatible mobile phones, close to the number of people who access the Internet by personal computer.[20]

The main attraction of the combined telephone and Internet service was that it was perfectly suited for Japanese life. Many Japanese people spend a good portion of their day away from home. It is not uncommon for people to travel two hours each way to work on a daily basis; many activities require long waits in line. The mobile Internet phone allows access to the Internet at any time from anywhere. (DoCoMo actually means 'anywhere' in Japanese. Ads also announce it as an abbreviation of 'Do communications over the mobile network'.) For commuters, the Internet is accessible from the train or bus, while walking or even riding a bike. Students waiting for classes, in the library or the cafeteria, salarymen walking to work, office ladies planning to meet friends for dinner, young couples searching for a movie, business people in urgent need of financial information, and countless others quickly recognized the value of the mobile Internet. By 2000, a full commuter train would often have a dozen or more people surfing the Internet. People riding bikes or walking along crowded urban streets can be seen working with their *keitai*. I-mode is particularly great for *hima otsubu* (literally 'crushing free time'); coffee shops are also filled with people typing away on their phones. I-mode users, therefore, have very different usage patterns from people who go online with their personal computers. While the average time spent online for a Japanese personal computer internet user is thirty minutes, the typical i-mode subscriber goes online for about two minutes. People will use i-mode for very short periods of time while waiting in line or commuting by train or bus.[21]

As a consequence of the exceptional popularity of i-mode, each passing month saw the introduction of new services. Japanese consumers (the data is almost exclusively in Japanese) now have rapid and cheap access to train schedules, restaurant menus, hotel and dinner reservation systems, taxi cabs, GPS services,[22] and numerous information sources, including news, sport scores (especially beloved sumo results), weather and traffic reports. Much of the information is targeted at specific age groups- there are a high percentage of young female users – and some content is free while other material is available for a fee. By 2000, several national magazines appeared catering to young *keitai* users, offering updates on new web-sites, downloadable games, music sites, and related information. Companies looked around, saw literally thousands of potential consumers using their *keitai* and raced to bring their services to market, thereby making the telephone terminals ever more useful and the service that much more cost effective.

Most striking has been the fact that the Japanese mobile internet has become popular with all age groups. There were almost as many users between 20 and 24, a critical age for technological adaptation, as there were above 40 years of age. The latter age group has often been very slow to adopt computer based technologies. One of the most unusual elements in the *keitai* revolution is the degree to which it has not been centred in a particularly age.

By the winter of 2002, subscribers to DoCoMo's i-mode had access to about 3,200 official I-mode sites and 60,000 unregulated internet sites (reached by typing in a URL). These official I-mode sites, accessible under the i-menu button, are closely monitored by DoCoMo to ensure that they are appropriate, interesting and easy to use . The i-menu is dynamic and constantly being updated. A 'What's New' section gives users information on the newest additions to the i-mode site and a simple mention here can send the number of hits a site receives skyrocketing. The official sites are by the far the most popular and DoCoMo is very very careful about which sites are approved. There are numerous content providers waiting for DoCoMo to approve their sites.[23] This is an interesting contrast to the usual view that the internet's strength is in the fact it cannot be controlled or regulated.

One service that did not initially meet DoCoMo's standards for safety and security was ImaHima ('Are you free now'?), a 'socializing and scheduling service enabling mobile customers to locate and contact friends, schedule parties and events, meet new people and find information and activities based on their current location via mobile phone'.[24] DoCoMo was worried about the potential for criminals to meet victims through the service and changes were eventually made to the system to satisfy this concern.

Japan's m-commerce demands and creates a different business model than other e-commerce retail and service initiatives around the world. An October 2001 *Economist* article summed this difference up nicely: 'Internet users expect things to be free, and are prepared to accept a certain degree of technological imperfection. Mobile users are accustomed to paying, but expect a far higher level of service and reliability in return.'[25] *Keitai*-based Internet services permit users to do much of what PC-based Internet users can do. Email, news, sports information, weather reports, movie listings, store addresses and the like are much the same, although the data has to be read on a much smaller screen. The wireless technology is very different, and much simpler than regular Internet systems. The new web-sites have few pictures and operate on the basis of easy to use menus; data can be entered into the phone via the telephone key pad, but the process is cumbersome and difficult. While this is something of a liability – PC users can navigate with a mouse, type in instructions, and respond to a wide variety of visual, text and auditory directions – simplicity has its advantages. Even very unsophisticated computer users can be operational within minutes on the basic mobile Internet system. In addition, input in Japanese requires significantly fewer key strokes than does input in English (because of the hiragana and katakana syllabaries.) A contest recently held in Tokyo found someone who could input 70 words a minute on a *keitai*![26] (In Japan, young people who are ardent users of the mobile internet are sometimes referred to as the 'thumb-tribe' 'due to their ability to type on a keypad with one thumb at a rate much faster than adults using PC keyboards and typing with ten fingers.')[27] The next stages of development may well tip the balance further in favour of the *keitai*. Voice email systems, which read out email messages and allow for spoken replies, are currently being introduced. Similarly, faster Internet speeds will enable *keitai* users to download more graphics, music and content-rich web-sites.

Where the mobile phone system has a particular advantage is in the charging system. E-commerce in Japan unfolded slowly, in significant part, due to the consumer resistance to providing confidential credit card information over the Internet. Further, credit card systems facilitated sizable purchases – books, electronics equipment, clothing, airfare, hotel reservations – but did not encourage micro-purchases, long known to be critical to the sale of content via the Internet. A simple example illustrates the point. Very few people have signed up to receive a full magazine subscription over the Internet; most of the publications that attempted to use this model experienced severe financial difficulties. But consumers might be very pleased to pay a small amount – one or two cents – to read a single article on a topic of personal interest. News providers can see the advantage in an arrangement whereby several tens of thousand of readers around the world might readily sign on to learn about the winners of the Oscars, current sports scores, the thoughts of a well-known columnist, or up to the minute news about a natural disaster, political scandal, war or business calamity. Current PC systems are not well-set up for such micropayment business activities (although various electronic wallets and charging systems are on the market).

The keitai, in contrast, arrived in Japan with the payment system in place. The same telephone companies that provide the service, having set up their accounting system to charge clients on a per packet basis, could easily extend their payment systems to incorporate the collection of revenue from content providers. (Of course, the telephone companies charged for this service, thus providing a revenue stream from both the consumers and the web-site companies.) Japanese users already have access to numerous micro-payment type products, with many more in the works. One of the simplest provides an excellent example of the market potential of mobile commerce. Bandai's Tare Panda, a noted Japanese animation firm, has the copyright to over tens of thousands of cartoon images. Shortly after the i-mode phone became popular, Bandai offered a very basic service: for as little as 100 yen per month (under $1 US) a subscribe could receive a new image for their *keitai* every day. The character would, for a 24 hour period, be comparable to the screensaver on a PC, visible whenever the Internet or telephone service is not in use. Bandai signed up 300,000 users within a few months, producing a US$10 million a year profit from content it already had in house. [28]

Club Giga's *chakumero* is another popular service that currently has about 100,000 subscribers. *Chakumero* are ring melodies, the little bits of a popular song or familiar jingle, that play when a phone rings. The first thing many new purchasers of an I-mode service do is download a melody for their phone. [29] (In fact, according to Infoseek Japan, *chakumero* is the most frequently used keyword search, followed by screen picture.) [30] The charge for these and other Internet services and purchases is simply added to the monthly mobile phone bill – which, incidentally, users can pay over their *keitai*. As *Wired* magazine pointed out, the range of personalized downloadable items expanded with the addition of animation:

> Hello Kitty, the adorable little pussy-cat that already adorns everything from bank cards to hot dogs, now appears on i-mode screens as well, chiming the hour and doing a little dance. J@pan Inc. magazine reports that 9 of the top 10 Java downloads on i-mode are games – everything from mah-jongg to *Shit Panic*, in which you try to catch the stuff as it falls and flush it down the toilet. But Mickey Mouse is big, too – and for just Y200 a month you can design a clock using any Disney character you want and make it your screensaver. 'When the clock strikes the hour, stars go off', says Mark Handler, who heads international operations for the Walt Disney Internet Group. 'It's exciting to watch'.[31]

Among the 60,000 non-official sites accessible with i-mode, many are abridged versions of regular web sites but others have been especially designed for the mobile environment. Not all of the content is serious. In fact, as one writer summed it up, 'Much of that content is about sex, sports, sex, astrology, sex, animation and sex.'[32] (The proliferation of Internet dating sites accessible through the *keitai* led to numerous complaints about solicitation attempts and extortion, robbery and rape. In December 2002, the National Police Agency announced a plan to ban the dating sites.[33]) Nonetheless, along with sex, sports, astrology and animation are the practical sites listing transportation routes and schedules, restaurant and concert information, updated sports scores, stock prices, short games, music clips, etc. There also many unique sites – a Starbucks locator service; a site 'that brokers deals between drivers and cargo companies'[34]; Photonet that lets subscribers deposit their personal photos which can then be accessed by prospective dates[35]; Warikan-kun 'where after-work groups of imbibing colleagues can easily calculate how much everybody has to pay – including a discount for people who came later (and presumably drank and ate less) and a proportionately higher charge for the boss'[36] – and many others.

Of i-mode's 1800 official sites, approximately 70 per cent are free and the remaining ones charge a fee. The free sites make money by selling products or services through their web-sites. Tsutaya, a big video and CD rental chain, earned more than 100 million yen per month in 2000. (Tsutaya also sends redeemable electronic coupons to its i-mode users. During these video rental promotions, rental receipts have jumped by 60 per cent.)[37] Some securities companies have made Y800 million yen a month. According to the President of DoCoMo, the fee-charging sites usually charge between 100 and 300 yen to subscribe. Most need about 10,000 subscribers to remain in business. While for some firms 10,000 subscribers is a big stretch, there are others with over 2 million subscribers, each paying about 200 yen a month resulting in monthly sales of about US$4 million! About half of all i-mode users pay for content and those users who pay subscribe to an average of 2.2 content sites.

The Japanese mobile internet system is succeeding because it is simple to use, portable and offers an easy payment system, making it very attractive to consumers. At the same time, both content providers and the parent company are able to make money. Content providers earn revenue through subscription fees and/or through the products or services they sell through their web-sites. DoCoMo makes money by charging a 9 per cent commission to the content providers and charging users monthly fees and data downloading costs. A typical i-mode user spends about 400 yen (US$3.25) a month on content subscriptions and 2,000 yen (US$17) on downloading content.

The combination of benefits to consumers, content providers and the parent companies created an overnight success. Usage soared and rival firms moved quickly into the market. What differentiates Japanese m-commerce from e-commerce models in the rest of the world is that its practical, easy to implement (the web-sites have to, for technological reasons, be simple and manageable) and, most importantly, profitable. E-commerce firms in North America and Europe continue to struggle, with significant bankruptcies becoming regular events; the promise and hype surrounding the new business model has not, in the retail and service sector at least, been matched with results. The opposite is true in Japan. The i-mode phone was introduced with less fanfare than might be expected, and within a few months, DoCoMO was having to restrict new service applications in order to keep up with demand. Predictions for mobile phone based e-commerce are that it will increase to 2.45 trillion yen (about US$20 billion) by 2005.[38] (DoCoMo benefited enormously from the surge in consumer interest in their product. The stock price soared. In October 1998, a single share of DoCoMo stock was worth

US$44,425, among the very highest in the world. The company's requests for additional capital found receptive audiences, and the firm soon dwarfed even its parent, NTT, in total market value.[39] As of October 2001, DoCoMo's share price has fallen to US$13,250, partly due to the five month delay in the launch of its third generation system. The stock subsequently fell much further, due to the general decline in technology stocks, increased competition for *keitai* markets in Japan, and weak returns on investments in overseas markets.)[40]

While at the end of 2001, DoCoMo was by far the dominant wireless carrier in Japan with over 60 per cent of the mobile internet market, it is not the only player in the game. Some of DoCoMo's competitors are also doing very well themselves. Number two in the market is KDDI's EZweb which had about 6.7 million subscribers as of the end of March 2001. EZweb is the only mobile internet service in Japan to use the Wireless Application Protocol (WAP). WAP is more difficult for developers to use so fewer sites have been created for it.[41] As a consequence, EZweb users have access to only 3,000–4,000 sites in comparison to i-mode's over 60,000. J-Phone's J-Sky service has slightly fewer subscribers than EZWeb, with 6.2 million users. J-Sky has a creative and innovative brand image as it attempts to come up with new services which will appeal to its primary target market, young users. J-Sky subscribers are offered about 650 official content sites and 6,000 unofficial ones and in addition, they are also able to access any of the unofficial i-mode sites.[42] The final two competitors are DDI Pocket's H″ service which has about 2.5 million subscribers and the Astel Group's Dot-I.

Japanese businesses are now working assiduously to identify future m-commerce possibilities. The national interest in geographical positioning systems – a real attraction in the maze-like urban centres – suggests a natural connection with mobile phone technology. Advertisers have developed systems – and in m-commerce, new product innovations spring to market on a weekly basis – so that consumers will receive a telephone message whenever they walk past a particular store, perhaps to be offered a discount if they enter immediately. (The pricing model on this service is quite simple. Keitai users who enabled their phones to receive such advertisements would be 'paid' for accepting advertisements; businesses would, in turn, benefit from having access to consumers who are immediately outside their premises.) Increased transmission speed will permit content providers to upgrade the quality of their sites and the range of their services.

Japan has an extensive network of convenience stores (seven-eleven, Family Mart, AM/PM) around the country. These are easily converted into delivery and pick-up centres for m-commerce purchases, thus adding further to the ease of digital shopping in the country. In larger Japanese cities like Tokyo and Osaka, almost everyone commutes by train and many millions of people pass through the larger stations each day. For this reason, a delivery and pick-up service at the main stations would be very convenient for the vast majority of urban residents. JR East, one of the railway companies, has therefore begun developing collection sites at many of its larger train stations, designed specifically to serve the needs of m-commerce and e-commerce purchasers. The viability of these systems in Japan can be sharply contrasted to North America where most people commute by car and where one would be hard pressed to think of either a series of central locations or a national chain of outlets that would be convenient for the majority of the population.

DoCoMo has also been working to target business users for its i-mode service. While the initial subscribers were individuals, now DoCoMo wants to encourage entire companies to subscribe to i-mode by showing them how it will help their company work more efficiently and effectively. On the DoCoMo web-site, one section entitled Business Success Stories gives examples of firms who have used i-mode to increase their business productivity. Illustrations include American Family Life Assurance Company (AFLAC) agents who carry i-mode phones so that they can easily retrieve the most up-to-date information from their company's database and show it to their clients, and Goldwin, a sportswear company, which has given its salespeople i-mode so that they can submit sales reports and customer feedback directly and immediately to the head office.[43] I-mode clearly has considerable potential for corporate consumers.

One of the main attractions of the mobile Internet is the potential it offers for drawing together information in a rapid and accessible fashion. A newly launched service in Kobe, near Osaka, provides a good indication of how the system works. City Walk is a community-based Internet system, designed to provide consumers with access to information about restaurants, events, services and shopping.[44] It is offered over both the regular Internet and the m-phone system. It is with the latter that the real advantages of City Walk become apparent. In understanding the attraction of the system, it is vital to remember that 78 per cent of the Japanese population lives in urban areas (compared with 41 per cent in the United States) and that Japanese cities are densely settled, eschew any kind of regular grid road system, and challenge the most intrepid orienteer.[45]

Consider the following illustration of how this works. Two friends, working at opposite sides of town, decide to meet for dinner. One searches the web, at home, office or by phone, and emails a list of selected restaurants (with clickable links) to her friend. They make a selection, by email or by phone. Both now have a map that shows them how to find the restaurant. One, viewing the restaurant site, can either press a button and speak to the restaurant or can email a reservation request. Having confirmed a place, she then marks on the map a public location near to a subway station and emails the map, with the site marked, to her friend. They now know where to meet. Once they have arrived at the chosen site, they can then use the *keitai* to guide them to the restaurant (instructions are incredibly precise, indicating the number of metres between key points, providing the names of stores and buildings en route and, with the right technology, using a GPS to allow the friends to mark their progress). This is not the only service. The retail arrangements are similar, in that the City Walk system permits users to contact a wide variety of stores. Goods can be ordered over the *keitai*, and the purchaser can indicate where they want them delivered, either to a home or to a nearby convenience story. Similarly, City Walk tracks over 2,000 events at a time (movies, concerts, sporting contests, cultural activities, festivals, public meetings, lectures, conventions, and the like) and includes a quick reservation and ticketing system.

J-Phone launched a location-based service in May 2000. J-Navi allows users to enter a phone number or address and then find the nearest subway station or particular restaurant or kind of shop in the area. Searching is free but there is a small charge to download the information and an additional charge if users wish to download a colour map. By its third day of operation, J-Navi was receiving 1.6 million hits per day.[46]

The attraction of the system is somewhat restricted to Japanese environments, or other heavily urbanized areas with confusing grid systems (which includes most Asian cities and many of the older European centres). The combination of maps and printed directions (confirmation of business meetings in Japan typically include a faxed map, often with detailed written directions) and, more recently, GPS systems, is an enormous advantage in Japan. The same difficulty with urban navigation lies behind the widespread adoption of GPS navigation systems for automobiles, now widely used in taxis and becoming more common in higher end private automobiles. Most North American cities, and those developed largely in the post-1960s period, have more easily navigable street systems. The m-phone directional service would be much less attractive in such centres.

While m-commerce is, therefore, ideally suited to the unique characteristics of the Japanese market, it is also an eminently exportable nexus of technology and service.[47] NTT DoCoMo has entered into market development agreements around the world, including an important alliance (minority equity partnership)with America On-Line (AOL) in the United States.[48] Similar relationships exist with Hutchinson in the United Kingdom and Hong Kong; KPM mobile in the Netherlands, Germany, Belgium, Hungary and the Ukraine; Telefonica in Brazil; and KG Telecom in Taiwan. DoCoMo's competitors are likewise signing up international partners in the hope of extending the Japanese m-commerce advantage overseas. The systems will face formidable competition in the USA, which remains saddled with regulatory complications in the mobile phone field and where vast distances undercut many of the i-mode phone advantages, and Scandinavia, led by Finland's Nokia company, which enjoys a very high rate of mobile phone use and is poised to switch to the Internet telephone. But in much of the rest of the world, the high cost of installing telephone and fibre optic lines into private homes gives the Japanese mobile Internet a formidable advantage. Not surprisingly, therefore, Japanese firms see Asian markets as a high priority and forecast substantial opportunities in most countries.

The world of m-commerce in Japan is expanding rapidly. In December 2000, DoCoMo launched new commercial services which allow access to movies, news and sports via mobile phones. DoCoMo announced that there would be 47 channels available with 130 sub-menus of content from providers.[49] Then, at the end of January 2001, DoCoMo launched its Java technology enabled I-mode phones.[50] (Two weeks later, sales were briefly suspended due to a software glitch; DoCoMo provides a good example of a firm that is able to introduce new products to market with considerable speed and not suffer unduly in terms of consumer dissatisfaction from service or technological problems.)[51] The addition of Java makes many more complicated functions like networked games, real time stock prices, chat software, business support programs and, possibly most importantly, software enabling secure mobile commerce transactions. The cost for this enhanced service is 200 yen a month, (US$1.80), 15 yen (13 cents) per minute and certain charges for the content.[52]

And this all appears to be just the beginning of a 3G transformation, of which Japan is an important international player.[53] DoCoMo and Matsushita Communications International (which makes the handsets) launched their third generation mobile phones in the Tokyo area and parts of surrounding prefectures in October 2001 and expanded the service to the Kansai and other regions of the country soon after. This 3G phone service, named FOMA (for 'freedom of multimedia access'), was initially scheduled to begin in May 2001 but a series of problems delayed the launch until October 2001. FOMA initially only included voice and data transmission but video and music distribution began in late 2002. By that point, however, DoCoMo had signed up only 140,000 subscribers, far less than the 1.38 million for which it had been striving.[54] The low numbers were due to the short battery lives of the handsets and the restricted service area. Both problems, officials believe, will be rectified shortly as new handsets are due out in the fall of 2003 and the user area for FOMA will be expanded to 97 per cent of the population by March 2004.[55] (In the summer of 2002, DoCoMo also began selling a personal digital assistant compatible with its 3G cell phone service.) J-Phone launched its 3G cell phones in December 2002 and its tie-up with Vodafone Group Plc will allow its 3G subscribers to use their handsets for second generation services overseas.[56] KDDI Corp took a different tactic with its 'au' cell phone service and released phones with geographical positioning systems as its third generation system.

Digital camera equipped cell phones capable of taking, sending and receiving photographs are another recent sales success. J-phone launched its first camera-equipped handsets at the end of 2000, KDDI entered the market in April 2002 followed by NTT DoCoMo's I-shot camera phone in June. Total sales of these camera phones soared to 15 million units in January 2003 which meant that fully 20 per cent of all cell phone and personal handy phone subscribers had phones equipped with a camera.[57]

To assist consumers overwhelmed by the increased complexity of the new phones, the various carriers offered classes to show people how to use the various functions these new phones offer. Some classes served absolute novices (an attempt by the companies to broaden their subscriber base by convincing less technologically interested or proficient people to give the *keitai* a chance) and focused on easy to use models. DoCoMo's easiest model line – the Raku Raku or Easy Easy phones –

has comparatively large characters and the layout has been specially designed to make the handset easy to use. One model gives voice guidance on how to use i-mode and it even reads e-mail messages aloud.[58]

The relationship between the mobile operators and the handset manufacturers in Japan has been of critical importance to the ability of Japanese firms to continue to launch new and innovative services. An October 2001 *Economist* survey of the mobile internet pointed out why:

> When a Japanese operator wants to launch a new service, such as picture messaging or videotelephony, it can specify in detail how that service will work, ask manufacturers to build the appropriate handsets, and ensure that these are available when the service is launched. Operators in other parts of the world are too numerous to be able to boss the handset makers around in this way. Instead they face a chicken-and-egg situation. There is no point in their launching an innovative service unless handsets to support it are available; but until such a service is launched, manufacturers have no reason to produce handsets to support it. This explains why, for example, handsets with colour screens are still a rarity in Europe, though they are commonplace in Japan. Colour screens make phones more expensive, and consumers will not pay for them unless there are compelling services that use colour. No such services exist, so nobody buys colour handsets, so there is no reason to develop colour services.

Ballooning research and development costs for 3G handsets led NTT DoCoMo to decide to provide 40 billion yen (about half the development costs) in subsidies to NEC Corp., Fujitsu Ltd. and Matsushita Electric Industrial Company for their new handsets. The development costs for the necessary software and microprocessors for the FOMA handsets are substantially higher than those for conventional cell phones. As one analyst observed, 'By shouldering half the development costs, DoCoMo not only aims to slash the prices of 3G handsets, but also hopes to gain joint patents and copyrights on software and hardware needed to provide new services and make the handsets lighter and longer-lived.'[59]

DoCoMo established Yokosuka Research Park, just south of Tokyo, to expand its research activities on new mobile phones. Almost all mobile phone makers (60 companies representing 90 per cent of the mobile phone market), with the exception of South Korean firms, set up operations in the park so that they can be close to DoCoMo.[60] The synergy between manufacturers and service providers – in mobile telephony as in other commercial areas – provides Japan with a distinct advantage, one that DoCoMo and other competitors have been determined to exploit.

Many possible future uses of the wireless technology include allowing mobile phones to be used to start rice cookers or heat up the *ofuro* (Japanese bath) from a distance – while on a commute home, for example. One new technology, particularly suited to senior citizens living alone and appropriate for Japan's rapidly aging population, enables the fridge or the hot water pot (a common feature of Japanese homes) to send a message to a mobile phone every time the appliance is used. If the fridge is not opened or hot water not used for a certain period of time, the appliance sends a message to the mobile phone. Thus alerted, the person then calls someone to check on the senior. All of these new technologies will be, of course, available for a fee. Companies are also developing electronic money, tickets and coupons, and establishing protocols to permit cell phone users to make purchases at vending machines or pay for parking lot services.[61] In January 2003, DoCoMo announced that it had developed the technology for a cell phone that can serve as an electronic wallet and event ticket: 'With data transmitted using the IC card's wireless communications function, the phone allows users to make on-the-spot payments at supermarkets by placing the handset near IC tags attached to products, for instance, sparing users the inconvenience of waiting in line at checkout counters.'[62] Cybird Company has developed a system that will enable cell phone users to access full panoramic images of different settings. This could be used by restaurants or amusement centers to show potential customers the length of lineups or by real estate agents or tourist areas.[63] Particularly exciting is the possibility mentioned by Madoko Tsutsumi, a technician at DoCoMo. As one journalist noted: 'Tsutsumi predicts that her customers will be able to consult a doctor and show the medic their injury on the phone-screen. "Say ahh", the doc will order and the *keitai*, held to the open throat will beam, like a camera, the image of an inflamed tonsil or a swollen gland to the medic's own screen.'[64]

Japanese companies are also exploring other potential uses of the wireless technology. Matsushita Electric Industrial Company and Finland's Nokia Corporation have agreed to jointly develop technology that allows data to be exchanged between all types of appliances. People would be able to remotely control audiovisual devices from their cell phones and send pictures from such phones without going through a computer. The two companies are also working on other services linking cell phones and home appliances.[65] Toshiba uses wireless technology for remote monitoring of various kinds of equipment like photocopiers, water filters and kitchen garbage processors, reducing servicing costs as repair people can be dispatched more quickly. NTT ME Corp. and AlphaOmega Soft Co. launched a new device which allows pet owners to feed and monitor their pets while away from home. IseePet consists of a water-and-food dispenser that can be manipulated from a personal computer or an internet capable cell phone, releasing the food when needed. A camera is also attached to the device so that owners can watch their pets.[66]

DoCoMo's president stated in a speech in June 2001 in Tokyo that he saw his company's potential market as much bigger than simply people. Wireless technology can be attached to everything from pets and cars to appliances and vending machines, he said. DoCoMo is, in fact, already involved with vending machine manufacturers and soft drink firms to enable vending machines to accept payments from wireless devices and use wireless technology to track inventory levels and monitor maintenance and transmit information on both to the companies involved.[67] NTT and the University of Tokyo, building on work started in the United States, have even begun research into the world beyond wireless communication – transmitting data signals through the body itself. At a fall 2002 demonstration at the NTT Atsugi R&D Centre in Japan, NTT and NTT DoCoMo demonstrated the technology. '... two people approached each other and slowly clasped hands. At that moment, data from personal digital assistants was exchanged. But this was not wireless communications. Instead, the data passed right across their hands, travelling through both bodies at broadband speeds of 10 mbps.'[68]

M-commerce is clearly only in its infancy in Japan, and as with the heavily hyped and somewhat postponed e-commerce (business to consumer) revolution in North America[69] and Europe,[70] considerable skepticism remains.[71] The massive adoption of the new technology, across age ranges and without reference to economic background or commercial activity, suggests that the m-commerce transformation is just beginning. The potential implications of this transition are considerable. A steady increase in *keitai* usage may encourage consumers to abandon their land-line, PC-based Internet connections. This, in turn, might actually create some real competition in the marketplace. PC sales are a critical part of the digital economy around the world; the much cheaper and easier to manufacture mobile phones are already attracting attention and money away from the standard-bearers of the e-economy. How this will affect sales, profits and future investment in PC research remains to be seen. The business implications of the new technology will obviously flow over into other commercial sectors. Although evidence in this regard is more anecdotal than systematic, the popularity of the *keitai* appears to be affecting other commercial activities directed at teenagers. Although the costs are comparatively small, widespread *keitai* use is slowly draining spending from other parts of the youth economy, as young people spend hours each month hooked up to friends, web-sites and commercial services. (At the Coming of Age Day holiday in 2001, a celebration of young people turning twenty, mayors in several cities left the festivities because the young people were too busy typing on their *keitais* to pay attention.)[72] It is not yet clear if the reduced spending will come in entertainment, clothing purchases, food, or other areas, but the expansion of m-commerce is certain to draw business away from other youth-oriented enterprises.

Early in 2001, NTT announced its intention to establish an L-mode system, patterned on the overwhelming success of the i-mode telephone. L-mode uses regular telephone lines to deliver high-speed Internet service to specially designed phones (with liquid crystal displays). Thus, Japanese consumers would be able to use the Internet without the expense of purchasing a personal computer. The potential of the service is clearly significant, although the planned summer 2001 roll-out was postponed while NTT East and NTT West worked out the regulatory problems relating to the distribution of the Internet over telephone lines.[73] By the winter of 2002, L-mode had over 200,000 subscribers, a rapid growth but a small portion of the much larger and faster growing i-mode market.[74]

The planned introduction of digital broadcasting in 2003 has likewise been touted as a potentially crucial development, promising to permit interactive communication through television sets.[75] Models of convergence in other industrialized nations focused on combining the Internet and television, with the telephone being a minor afterthought. The Japanese experience turns that on its head. The telephone becomes the pivotal element, with the Internet already connected and, soon enough, with television service linked up as well. Japan, incidentally, has for years produced an array of portable television sets, the smallest designed to be worn in place of a wristwatch. While there is some interest in portable TVs, market interest pales in comparison to that for the *keitai*. Portable televisions are rarely seen on commuter trains and subways cars, despite their obvious suitability for just such an environment. Keitai usage, after only a few years in the market, outnumbers both portable television and CD/cassette players combined.

What ultimately makes Japan's innovation so important is accessibility. M-commerce is cheap, easy and portable. It is affordable for young and old alike, and is remarkably easy to use. The mobile phones are obviously not site bound and hence can be used by commuters, advertisers, retailers and service providers in ways not anticipated by PC developers. Internet access while commuting, walking, sitting in a restaurant, or on a bike is vastly different than the PC-based standard that the rest of the world uses.[76] Because consumer demand is ahead of commercial promotions – the companies are starting to catch up,[77] but as consumer demand continues to grow, m-commerce will proceed in Japan with a more solid financial foundation than the e-commerce retail initiative has done in other parts of the world. Non-Asian nations remain singularly uninterested in Japan's *keitai* revolution – in large part because the major firms have invested heavily in other forms of wireless communication or Internet delivery and because the conditions that favour m-commerce in Japan are not readily apparent in most other countries. As the technology improves, retailing and service experience grows, and the integration of m-commerce into everyday life in Japan expands, perhaps the rest of the world will begin to take more notice.[78] The *keitai* revolution in Japan is, after all, perhaps one of the best implementations of digital technology for everyday consumers. For that reason alone, foreign companies, governments and consumers analysts would do well to pay attention.[79]

There is a salutary lesson in DoCoMo's remarkable success. Through 1999–2001, companies in Europe and North American bid aggressively for licences to offer mobile internet services in new markets. Prices skyrocketed as companies dreamed of matching or exceeding the Japanese experience. In the first months of 2001 as part of the general dot.com and internet collapse, mobile internet companies like Palm Pilot and Research in Motion endured major losses. They discovered that although technology is universal in its application, technology-based services have to resonate with local consumers. As of 2001, DoCoMo's system was the only commercially viable mobile internet service in the world. What flourishes in Japan does not necessarily work in the rest of the world. Applications of the Japanese model, or country-specific mobile Internet developments, have the potential to produce economically sound offerings (possibly through the application of 3-G technologies, which permit audio and video Internet services).

Mobile phoned based Internet access will remain popular in Japan as it fits so well with life in Japan. As of 31 March 2001, DoCoMo was the 11th largest company in the world and one of the major 'survivors' of the dot.com euphoria of the late 1990s. The company is extremely popular in Japan and is a market and innovation leader. Its most recent experiment, with 3-G telephones, was a less than admirable roll-out, and several South Korean firms have enjoyed equally or more successful audio-visual Internet implementations. There is also mounting evidence that, within Japan, DoCoMo may been nearing its saturation point. Between January and September 2001, the company saw a 52 per cent decline in profits, a logical consequence of the global dot.com bust, the inability to sustain break-neck growth patterns, losses from overseas investments and difficulties with its 3-G implementation.[80] DoCoMo remains at the forefront of mobile Internet research and is actively searching our new markets and applications, moving well beyond Internet capable phones targeted primarily at the youth market. What is clear is the company established a new technological platform for the Internet in Japan, and has sparked an m-commerce revolution that is one of the most significant and successful commercial applications of the Internet age. DoCoMo's determination to export its Japan-based success, through alliances in Asia, North America and Europe, raise the possibility that Japanese m-commerce may become an global standard in the years ahead.

4
Japanese E-Commerce

Beginning in the late 1990s, a dot.com 'revolution' swept through the industrialized world. Led by promoters such as Jeff Bezos, CEO of Amazon.com, Bill Gates, Mark Cuban and fueled by the most remarkable mobilization of risk capital in a century, the dot.com visionaries mapped out a strategy for the transformation of commercial enterprise. Massive 'communities' of customers would be carefully managed by loyalty-conscious companies. The ability to order a whole range of products, from music CDs to books to speciality foods and automobiles, would, they argued, destroy the bricks and mortar approach to retailing. As Internet use expanded, an entire generation of dot.com entrepreneurs scrambled on board, offering a full range of services, products and delivery systems, promising in the process to re-write the very fundamental rules of business. But not, it seemed, in Japan.[1]

The collapse of the dot-com boom in 2001/2002 has already blurred the hype and enthusiasm of the late 1990s. Billions of dollars in risk capital, particularly in the United States, home to close to 90 per cent of the world's commercial websites, fled from manufacturing and the service sector to dot.com companies. Teenagers and university students, blessed with entrepreneurial spirit and, in retrospect, very thin ideas, became multi-millionaires overnight. Investors emptied out pension accounts to support ill-formed ideas and to capitalize on the speculative frenzy that captured the western world. Technology companies, like Lucent, Nortel, Cisco, Sun, Oracle, Dell and others saw their stock price rise out of all of proportion to actual commercial value. But for a time, there appeared to be no end in sight. Business analysts spoke with unchallenged conviction about the 'new paradigm', and business schools rushed new programmes in e-commerce into service. Governments threw their weight behind the new model, sponsoring conferences, proposing new development programmes and industry incubators, and rushing legislation into place to match with the needs of Internet commerce.

Even though Japanese personal computer technology found receptive international markets, the business community lagged far behind the United States, Finland, Australia, Canada and other nations in experimenting with new business models. Fewer Japanese maintained personal Internet accounts and usage costs were, in comparative terms, oppressive. The quality of service lagged as well, as Japanese consumers rarely had access to the very best telecommunications solutions available on the global market. Poor and inconsistent connections to North America meant that Japanese Internet users did not have the same quality of service as American and other consumers had come to take for granted. Despite the (still strong) belief that the Internet would serve as a primary vehicle for the spread of English, making it both the language of the Net and by default of the world, the reality was that few Japanese consumers understood English well enough to use English language web sites with much regularity. So while Americans, Canadians, British, Australians, Scandinavians and many others rushed headlong into Internet commerce, Japanese followed some distance behind.

Of course, in a time when business leaders like Bill Gates wrote confidently of 'Internet speed' and cautioned that companies, and by extension, nations that missed out by weeks and months on the dot.com revolution were doomed for economic irrelevance, Japan's slowness off the mark seemed like a formidable problem. Comparative studies revealed that Japan lagged far behind in Internet usage and e-commerce implementations. The truth of the matter was that the country was actually only 18 months to two years behind the United States. Japanese entrepreneurs were slower than in other countries to launch new Internet initiatives, and investment capital came more slowly in Japan than in many other markets, but the distance between the leaders and the followers has, in the hysteria surrounding e-commerce, been greatly exaggerated. Japan was, by most measures, in the middle of the pack of industrialized nations in responding to Internet opportunities. In 1996, only 1,000 companies were conducting e-commerce consumer sales, a number that soared to almost 12,500 by the end of 1998. Only in the commercial chaos of the late 1990s did such a gap seem like an insurmountable barrier although, at the time, criticism of Japan's laggardly ways was pronounced. The passage of time would reveal that the delay in capitalizing on the Internet was not the fatal flaw so many foreigners predicted with such confidence.

At first, only a few Japanese companies ventured cautiously onto the Internet. Estimates in 2000 suggested that less than 0.2 per cent of consumer spending was done on the web – a sharp increase from the negligible amounts in 1997 and 1998. Many of the first major applications were foreign firms, localized for the Japanese market. Company managers enthusiastically predicted the immediate creation of truly global retail and service networks. Japan would, in this formulation, be part of an integrated, wired, Internet world. Japanese firms, in contrast, moved slowly. At first, several small retail outlets emerged to compete for the embryonic business to consumer trade. And then, as in North America, pornographers and gambling outlets discovered that guilty pleasures were among the most lucrative operations on the Net. (Japan's unusual pornography regulations, which prohibit the showing of pubic hair and genitalia and yet permit depiction of bondage, torture and other 'sexual' acts, made the country well-suited to the development of international pornography sites aimed at the Japanese market. Similarly, the country's well-known penchant for gambling, highlighted by the ubiquitous pachinko parlors, made gambling sites an obvious addition.)

Japan moved slowly, to the point where major Japanese business figures believed that the country was falling well behind its competitors. It was bad enough to lag behind the United States, the acknowledged world leader in e-commerce; discovering that other Asian nations, particularly Singapore and South Korea, were making great strides in capitalizing on the new technologies angered and worried national leaders. Hiroshi Araki, Advisor to the Electronic Commerce Promotion Council of Japan and Chair, Tokyo Electric Power Company, said of the country's performance:

> Following the United States, electronic commerce has finally reached critical mass in Japan. However, Japan's localized systems and commercial practices are major factors that hinder further development of electronic commerce, and the mature economic structure has brought the atmosphere that impedes structural reform. Today, countries in Southeast Asia invest in Internet technologies more aggressively than Japan does. Now our concern is that we are not only behind the United States, but also other Asian countries as well. To break through the current stagnation, we must change the corporate climates of the past with an aggressive spirit

seeking for innovation. Moreover, we are required to carry out this paradigm-shift at no less speed than the world.[2]

Major companies proved not to be the source of the inspiration and imagination Araki sought.

A much greater dose of imagination and synergy emerged from the two most potent forces in Japanese e-commerce – the short-lived vitality of Tokyo's Bit Valley and the expansion of the Softbank empire. Bit Valley, developed by Sony Koike and Kiyoshi Nishikawa in an attempt to stimulate an Internet revolution in Tokyo, brought together enthusiasts, developers and visionaries. Bit Valley's work was spurred on by the messianic Masayoshi Son, who threw his considerable weight behind the development of Japan's e-commerce capacity. Their efforts resulted in the concentration of e-commerce companies in the Shibuya district[3] (although the Bit Valley Association later defined Bit Valley to include all of Tokyo, given that many of its members were based elsewhere in the city). Bit Valley (Shibuya means Bitter Valley in Japanese) attempted to reproduce in Japan the energy, creativity and risk-taking of California's famed Silicon Valley and of the merger of entrepreneurs and Internet technologists in other cities and countries. Bit Valley shared many characteristics with its American counterpart, bringing together risk capital, youthful insights, and technological sophistication. Bit Valley became famous for its energy, enthusiasm and, equally, its parties, where Son made regular appearances to spur on the high tech developers. Despite the international preoccupation with Japan's economic woes, there was still investment capital to be found, although venture capital was much harder to come by than in North America. As one reporter observed of the development at its height in 1999:

> The organization's name is an abbreviated version of 'bitter valley,' the English translation of Shibuya, the hip Tokyo neighborhood where many of Japan's Netrepreneurs are headquartered. The group is starting something of a cultural revolution in Japan. In an economy long characterized by corporate behemoths and lifetime employment, a surprisingly favorable climate is emerging for nimble startups that aspire to become the next Yahoo! or eBay. As the Bit Valley group grows, its influence is spreading beyond Tokyo to the rest of Japan. Since its first meeting in March at a French-style cafe in Shibuya's Bunkamura theater and art center, Bit Valley's membership has swelled from 100 to more than 1,200, representing some

400 Shibuya Net businesses and the Japanese offices of Microsoft, America Online, and others. 'Japan desperately needs a break-through movement to make way for an Internet boom,' says Koike.

Bit Valley appeared to offer the perfect antidote for a recession-ridden Japan: risk capital, enthusiasm and opportunities for creative young Japanese looking to avoid a life as a salaryman. At their parties, the 'netrepreneurs' spoke enthusiastically about revolutionary changes in Japanese society and commerce, and about how the Internet was going to transform the way in which the country worked, studied, lived and entertained itself.

Company after company rushed to the fore, in a Japanese variant of the Silicon Valley boom which promised to revolutionize North American society.[4] Masayoshi Son's Softbank was in the lead, converting US dot.coms into Japanese offerings and investing millions in Japanese-based web-sites. The names and concepts shared much in common with developments in other countries and, as advertisers' commitments to the Internet reveal, a pattern of domination by a handful of familiar firms and considerable involvement from the financial sector was soon evident (Table 4.1). Indigo offered e-commerce services and brought the auction site Onsale to Japan. Cyber Agent promoted on-line advertising. Digital Garage worked with Infoseek for a time before changing into an e-commerce incubator. Rakuten's net shopping centre offered web-surfers access to hundreds of stores, including many major firms.[5] (By 2002, Rakuten provided access to 9,300 retailers, ran $64 million a month through its Internet

Table 4.1   **Top 10 web advertisers in Japan, February 2002**

Company	*Market reach(%)*
Amazon	65.0
Yahoo!	64.5
eBank	49
AXA Direct	40.4
Mobit	35.2
At-Loan	34.5
UFJ Bank	22.3
Japan Net Bank	21.1
GE Consumer Credit	17.2
AIC	16.9

Source: Nielsen/Net Ratings (www.netratings.co.jp)

site, and attracted some 18 million hits per day, making it the most popular Internet shopping site. As an example, the company offered between 60,000 and 70,000 varieties of wine.[6]) Netage started with car-sales and expanded into retail operations. Much of this activity was derivative, either following closely on American models or localizing a USA company's business model. But the energy was real, and hundreds of dot.com entrepreneurs rushed into Bit Valley to share in the excitement and the anticipated profits.

One man stood apart from the Internet mob. Masayoshi Son, child of Korean-born Japanese parents, possessed a passionate Internet vision for Japan, one which held the potential to launch the country to the forefront of the e-commerce world. Son, with a BA in economics from the University of California Berkeley and a patent for an electronic organizer (which he sold to Sharp), returned to Japan determined to participate in the digital revolution in the country of his birth. He established a software distribution company, Softbank, in 1981. Within a few years, this cultural outsider turned digital entrepreneur, had established himself as a major presence within the Japanese computer and internet sector.

Uncharacteristically for Japanese start-ups, Son and Softbank set their sights on the international marketplace. As early as 1996, he invested heavily in Ziff-Davis, an American computer magazine publisher. While this investment made a name for Son on the global scene, it proved to be a poor investment, as the firm was soon losing a great deal of money. Some stellar investing, including the purchase of close to 30 per cent of Yahoo! in 1996, however, gave Son the capital he required. This single investment, which cost a staggering $100 million, was worth a mind-boggling $8.4 billion in only three years and an astronomical $33 billion in February 2000. Son's well-publicized goal was to copy the pre-World War II *zaibatsu* model, creating an Internet behemoth which would introduce e-commerce to Japan and become the foundation for global leadership in the field. This dual goal – Japanese connectivity and global leadership – rested on Son's ability to mobilize Japanese capital, build viable markets for his e-commerce initiatives, and establish controlling positions with Internet firms around the world. Using his considerable personal stake in Yahoo! and other early Internet success stories as his investment capital, Son launched a remarkably ambitious acquisition program through his holding company, Softbank.[7]

Son's major Internet holding was Yahoo! Japan, the most visited site on the Japanese Internet and a major source of revenue for the company. He expanded into computer magazines and, on the heels of the dot.com expansion in the USA, into a series of Internet companies. By 1999, analysts estimated that Son controlled over 70 per cent of the Japanese Internet economy, through such companies as Yahoo! Japan, Forexbank (currency trades), E*Trade (stockbrokerage), E-shopping (toys) Japan, CarPoint Japan (auto sales), Onsale Japan (used car sales), GeoCities Japan (a portal), and Nasdaq Japan. Most of these companies were mirrors of North American firms, following known and potentially profitable marketing and business plans. Others, like Pasano Softbank, provided online employment services and, on the crest of the Internet wave, saw their market valuations skyrocket. Softbank held a substantial interest in almost all of these companies, and the dozens of others it soon added to its holdings. As one observer commented of Son in 1999, 'He is in a position to dominate the technology of the future in the world's second-largest economy, an awesome prospect. By extending its tentacles throughout Japan's Internet economy, Son's octopus looks ready to embrace the future – while much of corporate Japan is still at sea.'[8] By early 2000, Softbank was Japan's fifth largest company (measured in market capitalization).

Son did, indeed, intend to create a Japanese e-octopus, and promoted a 100-year plan (at some times, he is said to have spoken about 300 year time frames, just like Matsushita Konosuke, founder of Matsushita) for Japan's presence on the Internet. He approached investing in a straight-forward fashion: 'At its core, Son's plan is simple. The company takes minority stakes in companies that it thinks will be the winners in their business segments on the Web. In some cases, he tries to replicate an already well-established American brand overseas–such as Yahoo Japan, and E*Trade Japan. To give his portfolio companies an edge, Son has created what he calls a Netbatsu, a modern reprise of the old Japanese zaibatsu trading networks that swapped ideas, capital, and contacts to expand their industrial empires throughout Asia. Son's fund managers and the executives of his portfolio companies can share information and easily do deals together that benefit all parties.'[9] Following on the heels of his aggressive Japanese strategy, he began to invest heavily in international Internet initiatives. He watched Internet start-ups, looking for good ideas which had failed to reach their potential due to undercapitalization, poor implementation or weak leadership. Early in 2000, Softbank held equity positions in over 300 companies, including 160 in the USA, 120 in Japan and others in Asia and

Europe. He saw his Internet *zaibatsu* growing to some 800 Internet companies over the coming years, and believed that Softbank would become and remain a world leader in the field.

Son wanted representation in all regions of the world, believing that the integration of these companies would ensure Japan a major and continuing presence in the new economy. He held, further, that the conservative Japanese approach of both consumers and corporations threatened to leave the country out of this exciting and economy-changing transformation. Son was determined, singlehandedly if necessary, to break through the Nippon techno-phobia and launch Japan into the Internet age. As he said, 'Japan is going through the biggest social upheaval since the Meiji Restoration.'[10] For his efforts, he received both kudos and criticism:

> Son has been hailed by some as a change agent for the moribund Japanese economy, but critics wonder whether he's another cartel builder in a country full of them. HSBC Securities Inc. analyst J. Brian Waterhouse worries that Aozora Bank will turn out to be Son's 'high-tech piggy bank,' steering preferential lending to its partners. Similar questions have been raised about Softbank's controlling interest in Nasdaq Japan. The suggestion that he's not playing fair gets Son's back up. He insists both the bank and Nasdaq Japan won't unfairly favor his other companies. 'I shouldn't be criticized for doing the right thing for society,' he fumes.[11]

Like Son himself, the Internet community in Japan was young – uncharacteristically for Japan, many of the entrepreneurs were in their twenties or thirties – dynamic, fast-moving, risk-taking, and self-assured. They eschewed traditional Japanese management techniques and revelled in their contrariness. The young workers dressed casually, mingled easily with foreigners, and believed that their approach to economic and commercial development foreshadowed a new Japan. The booming opportunities for marketers, salespeople, and many other non-technical staff encouraged mid-career individuals to jump from their 'safe' jobs with big companies to the start-ups. As one commentator observers, 'In Japan, the Net has wrought far more than a gaggle of Web sites and dreams of dot-com dough. Net downturn or not, workers have been introduced to a new way of doing things – and they like it. Now, unwilling to slog through stifling 30-year career jobs, rising stars such as Iino, Takechi, and Nakagawa are heading for the caffeinated lifestyle of Web ventures. They're after new skills, more responsibility, and – maybe, just maybe – the chance to make it big at a cutting-edge outfit.'[12]

Like their American brethren, they believed uncritically in the inevitability of the 'new economy', although they expressed some puzzlement at Japan's slow adoption of the Internet. New companies exploded onto the scene, some to die a quick but exciting death, others to merge with competitors or to find corporate allies. A handful began to experience the commercial potential of the Internet. Programmers, web-designers and content providers/web editors were in short supply, with the start-ups offering high salaries to attract the 'talent', much of it from abroad. For Japanese workers, the draw was not the promise of high salaries and stock options – the magnets that worked so effectively in North America. Instead, Bit Valley and its counterparts offered Japanese workers a different employment model: 'A study by Unifi Network, a division of consultant PricewaterhouseCoopers, shows that Japanese overwhelmingly seek personal growth, new skills, and creative content when moving to Net upstarts. What happened to "show me the money"? Making oodles of moolah placed last on the list. Take Iino, the former banker and father of two, who went for a 65 per cent pay cut to become chief operating officer of Resonance, an online publisher and advertising agency. Despite the loss of income and stature, he could not be happier. He makes crucial business decisions daily, something he couldn't do at the bank. "I was on a smooth, straight path that anyone would understand, but I was living in the old paradigm," says Iino. "If I stayed at the bank, my path would have been static and linear."'[13]

Many new ventures exploded onto the scene, and their values soon matched North American dot.com stocks.. Hiroshi Mikitani's Rakuten, an online shopping mall offering hundreds of 'stores' (including access to several of Japan's elite companies, including Seibu), attracted a great deal of interest. Hikari Tsushin (the name means light-speed communications), combining mobile phone and Internet investment services, had in February 2000 a market value exceeding stalwarts like Sony, Honda or Matsushita. Hikarai Tsushin, like Trans Cosmos (with money in Amazon.com, Autobytel.com, DoubleCLick and Liquid Auto), invested heavily in other dot.coms and sought to challenge Softbank for Japanese supremacy. So, too, did Masatoshi Kumagai's interQ group. Foreign capital came in as well, drawing firms like J.H. Whitney (with over $100 million in funds for start-ups and other Internet ventures) and Mabrecht and Quist into the Japanese Internet field.

The speculative fever reached deep into Japan, although public enthusiasm never matched that in the United States. The Tokyo Stock Exchange created the 'Mothers' market – Market of the High Growth and Emerging Stocks – in the autumn of 1999 to service this demand and to enable companies to bring their firm to market much sooner than in the past. Son's Nasdaq Japan was established in an attempt to further 'democractize' share holdings in the Internet sector. New companies signed up, eager to tap into national and international markets for capital to fund their business plans. Liquid Audio offered Internet music services. Internet Research Institute provided consultancies. Met's Corporation designed software and Crayfish offered email connections.[14] Internet advertiser Cyber Agent likewise reached out to 'Mothers'. Brave forecasts of another 500 firms on the Mothers board by 2001 indicated something of the promoters' enthusiasm for e-commerce in Japan.[15] In the midst of euphoria over high valuations – Yahoo! Japan was in 1999 selling at 3,000 times earnings – some prescient warnings from Craig Nelson of Dresdner Kleinwort Benson: 'The mania for Internet stocks is looking more and more like a bubble. At the same time, the newspapers are announcing one after another venture fund being set up to invest in such stocks, including many who should not be dabbling in dangerous waters. It seems that even before Japan's metamorphosis into a nation of entrepreneurs has begun, it may be halted by the bursting bubble.'[16]

Most of the initial Japanese Internet ventures mirrored North American and other foreign initiatives. Son's Softbank empire was built largely around developing Japanese branches of many of the most important US web-sites. Small companies flooded onto the web, believing that instant access to millions of cash-rich consumers would result in quick profits. Like their North American counterparts, most were soon disappointed. DealTime Japan, supported by Omron, Mitsui & Co. and Credit Saison, opened operations in 2000, offering consumers access to over 200,000 products from almost 350 companies. The main feature of DealTime, an American business model brought to Japan, was the opportunity it provided for comparison shopping. The company was far from alone in the Japanese market, having to compete with such firms as Kakaku.com, EasySeek.com and BargainNews.net.[17] Priceline.com entered the Japanese market in 2001. Aggressive competition was not uncommon, convincing e-Bay to announce in 2002 that it was closing its Japanese operations.[18] As in North America, the Internet promised to revolutionize the book industry. Japanese firms, led by Kinokuniya Book Web, Kuroneko-Yamato's

Book Service and Yaesu Book Centre, followed Amazon.com's model, creating on-line search engines, ordering services and customer support and information. While online bookstores proved reasonably success-ful in Japan, they were not the 'category-killers' that Amazon.com and BarnesandNoble.com were in the United States.

Japanese companies put a great deal of effort into the creation of Internet 'malls', aggregations of large and small companies and avail-able through a single portal. The early leaders in this substantial sector came through web-specific ventures. These internet-sensitive firms knew how to capitalize on the technology's potential and were eager to try new approaches. Rakuten Ichiba, the most successful of the web-malls, for example, pioneered the use of email to stay in touch with customers. Yahoo! Shopping capitalized on its connection to the country's most popular web-site. Other web-malls included Bargain America, O-Kini-City and Umaimono Kai. Major corporations estab-lished Internet malls (NTT – Machiko, ASCII – ARCS, Toshiba – C-Mall Cyber Wing Club, Toppan Printing – Cyber Publishing, Dai Nippon Printing – EC Galaxy), but they had trouble making headway against the faster, more innovative new companies. A third group of web-malls operated by Internet Service Providers, like AOL, Hi-Ho, Nifty, Plala and JustNet, drew on their connections to their customers and found significant market niches.

(The founder of Hikari Tsushin, Yasumitsu Shigeta, capitalized on the growing market for mobile phones and established a nation-wide network of phone stores. He then used the capital from this profitable business to invest in the enterprises and, like Son, tied his company's future to the development of web strategies. Shigeta's personal wealth shot into the stratosphere. He was worth close to $25 billion, and rivaled Son for the status as the richest man in Japan. Like Son, he invested heavily outside Japan, and by 2000 had 70 Internet ventures in Japan, China and the USA. Importantly, Shigeta sat on the board of directors of Son's Softbank Company.[19])

As e-commerce emerged with a Japanese face, foreigner observers were impressed with what they saw. As *Fortune* magazine reported in February 2000, 'Now that's a New Economy. Trouble is, just about everybody else in the world, mesmerized by path-breaking deals in America and accustomed to Japan's fusty establishment, thinks that Japan's e-renaissance is either unimaginable or irrelevant. They should think again.'[20] *Business Week* was just as enthusiastic: 'Japan's Net builders believe they can reverse the slide. And as the revolution they started gathers steam, more men – and women, too – will find inspiration in their success. That should further undercut Japan's ossified business structures and breathe more life into the manifold Net Ventures. At this early stage, Japan is experiencing entrepreneurial drive it has not seen for 125 years. ... If the Net builders keep pushing, the bullish prophecies of the Roaring Eighties could come roaring back: The 21st Century may be Japan's century after all.'[21] Journalistic enthusiasm for Japan's Internet matched the uncritical and less than prescient reportage that surrounded the dot.com explosion in North America and elsewhere. For Japan, however, it was accompanied by a pattern of post-bubble era misrepresentation and misunderstanding of the strengths and weaknesses of the Japanese economy.

The larger, more conservative firms did not simply stand still, despite Son's view of them as lethargic and unimaginative. Many of the major manufacturers made substantial investments in business to business e-commerce implementations. Others, particularly Fujitsu,[22] Sony, and Tokyo Electric Power (which invested heavily in fiber optic cabling) recognized the potential value of the new business model. Unlike the dot.coms, they could trade on decades-old name recognition and tradition, important values for Japanese consumers. And while they were not driven into the field by the same sense of urgency or with the same verve and creativity, they found respectable places for themselves on the Internet landscape. Under the direction of President Naoyuki Akikusa, Fujitsu shifted its corporate operations toward Internet software and services, focusing particularly on ISPs and Internet content. The firm bought up major ISPs, including Nifty Serve (the country's largest in 1999) and made major investments in digital imaging. The company later cooperated with Sakura Bank to provide on-line financial services and is working on expanding its web-presence dramatically.[23]

Some foreign firms, still small in number and impact, have made major commitments to Japan. Foreign companies have made critical advances on the dot.com front. Yahoo Japan, Amazon.com, and General Motors are examples of firms which tried, with some success, to establish a foothold in Japan. Amazon.com opened Japanese operations in the late fall of 2000, setting up a distribution centre in Ichikawa, Chiba prefecture and a customer service office in Sapporo. Interestingly, Amazon was prevented by Japanese regulations from discounting its Japan-produced books, thus denying the company its major competitive advantage.[24] Yahoo! Japan, the cornerstone of the Softbank empire, is consistently the top-ranked Internet site in the country . The firm produces almost $8.2 million per month in revenue.

One of its key services is e-auctions, which gave the suddenly price conscious Japanese an opportunity to buy and sell second-hand products. Yahoo! Japan launched its auction service shortly before E-Bay arrived in the country, largely cornering the market in the process.[25] The launch of a new foreign-owned e-commerce product, like the 2000 opening of Amazon.com's Japanese branch, were widely promoted, often with the implicit suggestion that these (largely American) firms were destined to break the Japanese log-jam in e-commerce services. Their experience, not surprisingly, has differed little from Japanese firms, marked by high initial expectations, faltering revenues, and dismal profits.

(North American firms, incidentally, paid little attention to the net-based opportunities in Japan. According to a report from the Building2Information Group, almost exactly half of the United States's top 100 Internet retailers offer no sales to Japan. Among those that do encourage sales to Japan, few make much of an effort. Only 17 per cent of the top 100 internet retailers in the United States offered Japanese language websites.[26] The situation in Canada is much the same. A 2000 survey of Canadian firms who claimed to be committed to the Japanese market revealed that only a tiny number had any Japanese language information on their sites.)

As in North America, the hysteria surrounding e-commerce blinded business people and investors to the need for sustainable and identifiable cash flow. Expectations of immediate profits were set aside in the belief that long-term benefits would justify's investors's confidence in the commercial concepts. Many believed, in Japan and elsewhere, that Internet advertising would buttress e-commerce implementations. Japan's two largest advertising agencies, Dentsu and Hakuhodo, were early converts to Internet advertising , creating subsidiaries to work on the new market. While the initial response was tepid, promoters remained optimistic about future returns. Dai Ichi Kikaku differed from the others in using the Internet to promote international clients, particularly household products. Sony transformed the newly purchased Tokyu Agency International into Intervision and focused a significant part of its effort on the Internet and related digital technologies.[27] The early success of Yahoo! Japan, which produced rapid profits based on advertising revenue, spurred others to place their confidence in banner and click-through advertisements. As elsewhere, the promised Internet advertising revenues failed to materialize. Consumers remained loyal to newspapers, television and other promotions, and did not turn to the Internet with the alacrity anticipated.

Also as happened elsewhere, companies discovered that Internet information – details on products, prices and services – and not advertising was playing an increasingly prominent role in consumer's purchasing decisions.[28]

The excitement and high stock valuations lasted only a short time. Bit Valley and Son's plan for Japan's emergence as an Internet power imploded, about half a year before the dot.com crash in North America. The famed parties and high energy that suffused the enterprise for half a decade dissipated in an avalanche of systematic bad news. What the self-assured entrepreneurs discovered – and what the American entrepreneurs encountered a short time later – is that the euphoria of the IPO (initial purchase offer) and the ease with which money was raised initially was no substitute for solid business plans. The crash was smaller in Japan than the US for fairly basic reasons: 'The systems technology lag and the fact that venture capitalism in Japan is still relatively small fry compared to the US meant that the Internet bubble came here later than the US, and lasted less than a year, while the US lunacy of firing millions of dollars at online ideas lasted three years.'[29]

As in other parts of the world, there were few Japanese cheering on the often arrogant, techno-geeks who had rocked to fortune through the initial enthusiasm for the Internet and there were few lamentations when the companies and the lifestyle started to go under. Japanese consumers did not rush to the web-sites in numbers approximating those in other industrialized nations – and even the latter enthusiasm was not strong enough to buttress the flimsy cash flow experienced by most dot.coms. One after another, the Japanese dot.coms flared out, comet-like apparitions on the Japanese commercial landscape. They did not all disappear. A small number of Japanese web-companies soldiered on, generating sufficient cash flow to keep the corporate doors open. Some of the more popular sites were, indeed, foreign-owned or foreign-connected companies, like Amazon.co.jp and Yahoo.co.jp, which believed that Japan had considerable long-term potential.

In an Internet world full of audacious business plans and grandiose visions of the digital future, few countries had an individual with the complex ambitions of Masayoshi Son. He was determined to bring the Internet to Japan, to introduce e-commerce almost overnight and to expand across the retail and service sectors. Perhaps most importantly, to use income and capital from Japan to develop a Japanese Internet presence on a global scale. And for a short while, it appeared as though he would succeed. Son was one of the richest men in the world, the paper value of his many Internet holdings rode the dot.com roller-coaster to the apex (See Table 4.2 for an indication of Softbank's holdings). Just as quickly, Softbank collapsed in value when the Internet bubble burst, stripping Son and his companies of billions of dollars in value within a matter of months.

Son was not easily deterred. As one analyst wrote of Son, 'Indeed, Chairman Masayoshi Son seems to be the last true believer in the Internet Revolution. He already has invested $8.8 billion in corporate and venture capital money in more than 600 companies – about half of the dough coming from institutional investors. Now, he's got another $4 billion he raised before the market swoon that he plans on investing over the next two to three years. The goal is 800 Net companies in his portfolio. To get there, Son is now focusing on what he thinks are the hottest technologies showing the best growth prospects. That means networking, software-infrastructure players, and wireless-communications companies – as opposed to all but the most proven consumer sites.'[30] In Japan, even as the Internet boom was imploding, he was rolling out Yahoo! BB, a low.cost, high speed Internet service and talking enthusiastically about the future of the Internet in Japan. As Son reflected in December 2000, 'The market became filled with people who had no understanding of the Internet. ... They had no passion for the Internet, they just had a passion for making money. It was a boom without belief. With the market correction, those who are passionate just about making money are leaving. Now it is a better situation in which to make investments. Prices are more realistic.'[31] His optimism attracted a fair share of critics, who thought him unrealistic and ill-prepared for the swings in the marketplace. But Son was not deterred: 'The market goes up and down, you know. It's like the weather. Do you decide to become a golfer because today is a sunny day and then decide not to become a golfer because the next day is rainy.'[32]

Most of Softbank's over 600 companies were losing money and questions circulated about the future of the firm that laid the foundations for Japan's e-commerce market. As of the end of 2001, the company's value had collapsed from close to $200 billion to a still impressive but less lofty $27 billion. So convinced was Son of the open-ended expansion of the Internet that he had invested incautiously in dot.coms, and now saw major parts of his empire, including American firms, Buy.com and Webvan, struggle for survival. Cash flow problems and doubts about the value attached to Softbank's global holdings brought the financial vultures onto the scene. Aware of the criticism and scrutiny,

Son and his colleagues started closing companies that they deemed to be long-term money losers. Investments internationally were frozen, particularly in Latin American and Europe, and the company took a more cautious approach to the once open-ended Internet business.[33]

Son, himself, stayed optimistic, despite his ride on 'one of the shortest booms and busts in financial history – shorter and more brutal even than the dot.com rollercoaster in America and Europe. Since Japan's Internet bubble burst in March, the entire stockmarket has been reeling.' His personal wealth plummeted from some $68 billion – on track to overtake Bill Gates of Microsoft – to a 'mere' $2.7 billion. While his ambition remained, reigned in by financial difficulties, he shifted his attentions to the development of a high speed Internet service. Softbank consolidated its European offices in London, closing operations in Paris and Munich, and retrenched its activities in other parts of the world. At much the same time, however, Softbank expanded its operations in Asia, combining with the World Bank to support Internet companies in developing countries in Southeast Asia.[35] Once a high-profile promoter of Japan's Internet future, and a frequent speaker at formal and informal gatherings, Son assumed a much lower profile, seeking to bring his once high-flying firm into profitability.[36]

Son's gamble on cheap broadband Internet service was just that. Many observers believed that the Softbank's very future was at stake. The hardware investment, promising to challenge NTT's domination of Internet services by offering low cost, reliable, and very fast Internet connections, seemed like several steps backward for the firm and

Table 4.2 **Softbank Holdings (selected), September 2000[34]**

Japan: Softbank Finance, E-Trade Japan, E-Trade Securities, Morningstar Japan, Softbank Investment, Softbank E-Commerce, CarPoint Japan, E-Shopping, Priceline.co.jp, Softbank Commerce, Softbank Media and Marketing, Softbank Broadmedia, Yahoo! Japan, Cisco Japan, Nasdaq Japan.

United States; E-Trade, Net2Phone, PeoplePC, UT Starcom, Yahoo, ZDNet, Softbank Venture Capital, BlueLightc,om, Toysrus.com, More.com, Softbank Capital Partners, Kozmo.com, Naviant, Smart Age

Asia: Yahoo! Korea, Alibaba.com, Softbank China

Europe: Yahoo! UK, Yahoo France, Yahoo! Germany

Other: Softbank Latin American Ventures, Softbank Emerging Markets executive who promised to lead Japan to global Internet prominence. Son remained convinced, however, that handsome returns could be still found in the Internet world, even through such as yet unproven sectors as advertising and content sales. Son was not worried about short-term returns:

> I'm not much worried about that. It's a 100-year revolution. As it was for the auto industry, or the electronics industry, or the telephone industry. Narrowband is only the very beginning of the true technological depth that the Internet can deliver. What kind of content can you deliver on narrowband? Do you get excited about text and still pictures? Do you cry over them? Broadband will enlarge the total market size. It is going to open up all kinds of new opportunities. The scale will be hundreds of times bigger. People will start paying for content.

He intended Softbank to be there, and to be in a position to make sizeable returns, when people started to pay.

The most public manifestations of the Internet hysteria in Japan – the Bit Valley Association and its famed parties – all but disappeared. Where hundreds used to gather to listen to words of wisdom from Son and others, companies closed, offices stood empty and visions of Internet transformations were dashed. Softbank lost more than 90 per cent of its value. High-flyer Hikari Tsushin collapsed in a flurry of reports about shoddy financial reports and management. Sony Koike, whose consultancy firm and incubator company Netyear, had played a major role in getting Bit Valley off the ground, managed to stay afloat. As the sector reeled in chaos, Koike's company survived, if it did not flourish as before. By following a more moderate course than Son and Softbank, Koike was able to pick up promising companies, ideas and talent, and shifted gears away from the moribund Internet market and toward the exciting mobile commerce field. [37]

Softbank continues, and most of its Japanese Internet companies remain in operation. Son is no longer the superstar of the Japanese economy that he was in 1999/2000, and more than a few Japanese business leaders took satisfaction in his fall from pre-eminence. His critique of Japanese conservatism and his bold and unreachable ambitions for Japan's Internet economy earned him some friends, many followers and a legion of doubters. The dot.com bust appeared to prove the wisdom of the critics, but Masayoshi Son is far from finished. Even his much-reduced e-commerce empire still leaves him with hundreds of millions of dollars in assets. He and his companies will remain major players in the Japanese Internet economy.

Japan had not been bitten as severely by the dot.com bug as North America and parts of Europe, but it was close. The country had ready risk capital, the same kind of young web-entrepreneurs that emerged in the USA, and new, business-friendly stock market regulatory systems. Over 100 firms issued IPOs (initial placement offerings) in 1999, double that the following year and, before the boom came off the Internet rose late in 2000, expectations were that the surge in web-based commerce would continue to sky-rocket.[38] Closer examination reveals the preponderance of Softbank and Son's various enterprises. Hindsight indicates that few of the Internet companies had the depth or business plans necessary to sustain their high valuations. The firms that did start up tended to be smaller and less grandiose than the pets.com, etoys.com and grocery delivery businesses that promised North American investors imminent domination of entire market sectors, but there were hundreds of them ready to capitalize on the new opportunities. For several years, as stock prices in the United States soared – and languished in Japan – and as anecdotal evidence mounted that a fundamental restructuring of retail sales was under-way, foreign observers mocked Japan's mediocre e-commerce initiatives. The subsequent collapse of the dot.coms in the United States gave Japanese entrepreneurs and investors an opportunity to gloat about the wisdom of their more cautious approach to digital business, although they wisely did most of this in private.

Japan, much like the rest of Asia, was clearly not well-suited for the first wave of e-commerce.[39] The country's computer and Internet infra-structure could not support nation-wide roll outs of new businesses. Too few Japanese homes had Internet connections, and too few residents were interested in using web-based companies. There was a problem, too, with payment systems. The Japanese consumers' notorious reluctance to rely on credit cards was compounded by a widely-shared reluctance to rely on Internet credit transactions.[40] Add to this additional factors a downturn in retail sales associated with the economic slowdown in Japan, traditional shopping patterns which high-lighted customer loyalty and local purchases, limited e-commerce web-sites in Japanese, and a restricted range of products available through web-stores – and the inability of Japanese firms to match the euphoria that surrounded the dot.coms in much of the western world before 2001 makes more sense. By almost all indicators, Japan did not produce the consumer enthusiasm, corporate creativity, and

commercial innovation of many other countries – and nor, as a consequence, did the nation's businesses endure the same shocking jolt that
occurred in the dot.com meltdown in North America. Even the
country's best and biggest companies were laggards in terms of the
Internet. A 1999 survey of the Internet sites of the world's top 100 corporations listed only one Japanese company, Sony (12th), in the top 20
web-sites. While Japanese firms represented the top five in Asia, the
results were far from impressive: the other four after Sony were
Mitsubishi Electric (24), NEC (26), Matsushita Electric (34) and Japan
Airlines (38). [41]

The early experience with Internet-enabled commerce, however,
pointed to several very promising developments. While North
American e-commerce foundered due to unreasonable expectations,
uncontainable costs, and technological challenges, the system exposed
a fundamental problem in the delivery field. Work patterns in North
America meant that homes were often unoccupied during the day,
making delivery scheduling difficult. In Japan, many adult women
with families remain at home, meaning that many are available
throughout the day to receive deliveries. While American firms vied for
control of the lucrative e-commerce delivery market – companies com-
peted to be seen as 'The Courier for the Internet' – the national and
international infrastructure was not well developed, particularly in
contrast to Japan's efficient, fast and comparatively inexpensive
system. Other elements – strong customer loyalty to major stores,
Japan-specific tastes and requirements, the low cost (compared to the
same goods provided through local stores) of many imported items –
auger well for the long-term future of e-commerce. As the previous
chapter argues, m-commerce (mobile) capitalized on these opportunities more efficiently and successfully than did the PC-based e-commerce model adopted in all other countries. This said, the foundation
for the long-term expansion of e-commerce appeared to be well-set. It
is worth noting that the early Japanese usage of e-commerce differed
substantially from that in the United States, with greater emphasis on
personal computer sales, much less on automobiles, and little involvement from the financial sector (see Table 4.3).

One of the long-standing barriers to e-commerce expansion in Japan is
the comparatively low rate of credit card usage in the country and the
unwillingness to use credit cards for on-line purchases. The nation's
superb delivery system helped address this problem, and allowed purchasers to pay upon delivery. Many Japanese pay their bills through
bank and postal account transfers, a system which adds to the complex-

Table 4.3 Business to Consumer (B2C) e-commerce, Japan and the US compared, 1998

Japan (% of market)	U.S.A. (% of market)
Personal Computers – 39	Auto – 43
Travel – 12	Travel – 16
Apparel – 11	Finance – 12
Books/CDs – 6	Personal Computers – 9
Food – 6	Books/CDs – 6

Source: MITI/Andersen Consulting Joint Research.

ity of an on-line transaction. The ubiquitous convenience stores have proven to be a critical element in the e-commerce equation.[42] By the summer of 2002, e-commerce collections and payments at convenience stores accounted for 10 per cent (170 billion yen) of all e-commerce transactions; cash on delivery, in contrast, had fallen from 50 per cent of all on-line sales to 30 per cent between 1999 and 2002. The fact that the stores are conveniently located and open around the clock make them a superb adjunct to the e-commerce enterprise in Japan.[43]

Japan was, by 2001, the second largest e-commerce market in the world, second only to the United States, and by far and away the leader in Asia (see Table 4.4). The country accounted for 70 per cent of Asia's e-commerce activity. Business to business e-commerce outstripped business to consumer transactions by more than 10 times, a disparity that is expected to decline gradually in the coming years. Within the business sector, automobiles and real estate together accounted for 50 per cent of Japanese e-commerce. Estimates made in 2000, before the dot.com burst took some of the enthusiasm out of the marketplace, forecast that Japanese Business to Business (B2B) transactions would amount to $180.2 billion by 2005, with B2C revenues lagging at around $24.6 billion.[44] The B2B estimates appear reasonably on course,

Table 4.4 B2C e-commerce, selected countries

Country	*Transaction values, 1999 (US$ 000)*
United States	24,170,000
Japan	1,648,000
Germany	1,199,000
United Kingdom	1,040,000
Canada	744,000
France	345,000
Sweden	232,000

Source: OECD Economic Outlook, 2000.

despite the collapse of Internet euphoria, and the B2C numbers have likely been advanced somewhat by the success of mobile commerce in Japan. Globally, Japanese and Asia e-commerce activity has fallen short of other industrialized nations. In 2000, Asia was responsible for 14 per cent of global e-commerce; that number is expected to fall to 10 per cent by 2004.[45]

Compared to other nations, Japan has been running in the middle of the pack, but far ahead of other Asian countries. (see Table 4.5). A major investigation of the 'e-readiness' of countries around the world conducted by The Economist Intelligence Unit concluded that Japan was 18th and belonged in a group it described as 'e-commerce contenders.' (The report described contenders as countries which 'have both a satisfactory infrastructure and a good business environment. But parts of the e-business equation are lacking.') This ranking left Japan well behind the USA, Australia and the United Kingdom, the top three nations, but also significantly behind Singapore and Hong Kong, and one spot behind Taiwan.[46]

Japan made substantial investments in the future of e-commerce, with the benefits and impact of these developments to be seen in the coming decades. Major corporations invested heavily in e-commerce research and government–industry projects have flourished across the country.[47] As with the Internet generally, Japan continues to move more slowly and methodically, and has avoided the Internet panic which has inflected many other governments and business communities.[48] There are significant efforts underway to develop new products and services, and a firm belief in the commercial inevitability of certain, unspecified Internet applications. In 2000, a consortia of major Japanese firms, Sony, Matsushita and Toshiba, announced the launch

Table 4.5   **B2B e-commerce revenues in Asia, 2000 and 2004 (forecast) (% of Asian total)**

Country	2000	2004
Singapore	1.4	2.2
Hong Kong	1.7	2.7
Korea	4.8	4.7
Taiwan	5.9	6.2
Australia	6.3	8.2
China	2.2	7.3
Japan	69.5	60.0
Other	8.2	8.7

Source: *Marketer*, 2001: http://live.emarketer.com/

of interactive services via digital television, a merger of television and Internet technologies. The creation of an expansive e-commerce plat-form promised consumers entertainment on-demand, interactive com-puter games, and tele-shopping. The initiative, which built off the country's commitment to digital broadcasting, brought together bitter rivals Matshushita and Sony. Said Tamotsu Iba, Sony Vice-Chair, of the partnership: 'The relationship between Sony and Matsushita is not as bad as people think. When it comes to the era of e-business, we need to ensure open platforms and technology. And customers would be the biggest victim and demand would be stagnant if we were to adopt sep-arate standards for upcoming digital broadcasting.'[49]

Consider, as well, an ambitious plan undertaken by Matsushita. The company's product developers have targeted web-based implementations as key to their future. To this end, and in partnership with NTT, Matsushita started to wire the city of Kanazawa, creating a huge 'digital laboratory'. The combination of city-wide fibre optic cabling and the pro-duction of specialized and coordinated networking devices is intended to provide a prototype for the Internet lifestyle of the near future. Matshushita, for its part, will monitor the citizens throughout their private and business activities, providing data for their product develop-ers. Through such divisions as their Multimedia Life Creation Centre, the firm is preparing Internet products for the very young and very old, pre-saging a web-based environment without keyboards, complex instruc-tions and significant technology proficiency requirements.[50]

Other major corporations, including Mitsubishi, made significant shifts toward e-commerce implementations. Driven by competitive forces in Japan and globally, Mitsubishi placed increasing emphasis on the Internet as a means of responding to the changing business environ-ment. The trading company, which for years flourished as an intermedi-ary, is finding that the Internet allows it to perform much the same function, but in a very different way. The firm manages supply-chain activities for other companies, including retailer Uniqlo. It established eMaterialAuction to facilitate trade in steel, and uses eMerchantBank to underwrite some of this activity. The firm intends to extend the model into other sectors of the economy. That Mitsubishu and other trading houses, like Itochu, are well-established internationally has encouraged the companies to capitalize on the Internet to take advantage of global competition and supply opportunities.[51]

Importantly, as was discussed earlier, however, the national govern-ment has not yet utilized its considerable financial power to drive e-commerce developments through massive investments in e-government.

Unlike South Korea, which used its government procurement system to require major suppliers to move on-line aggressively, Japanese administrative use of e-commerce models lagged even behind the private sector. The results for Korea are now clear: a higher percentage of government offices, corporations and private citizens online and a flourishing e-commerce marketplace in the nation. It is not that digital business has not attracted government support in Japan. But that encouragement has been of a traditional sort: incubators for e-commerce companies (including several successful initiatives in Osaka and Kyoto), government subsidies for businesses, and official support for infrastructure development. What the Japanese government has not done is lead by example or utilize its high volume spending power to spark an expansion in e-commerce. That, too, is intended to change through the IT initiative, but the delay has resulted in Japan's e-commerce initiatives lagging behind other nations.

The challenges associates with the Y2K software question also revealed important differences in Japan's response to the possibilities of e-commerce. The country was very late into the game of responding to the threat of the year 2000 crisis, and spent much less than other nations in retooling their operations to prevent a potential catastrophe. The calm response – a sharp contrast to the near-panic which set in in other parts of the world – had several origins. The reliance on legacy systems meant that they could be readily adapted to address the Y2K problem. Long-term employee loyalty meant that, unlike in the USA, many of the people who wrote the original programmes were still associated with the firm, thus saving expensive consultant's fees. There was also considerable reliance in Japan on the Imperial calendar (years are named according to the length of the Emperor's tenure on the throne, thereby eliminating the year 2000 as a critical element) within computer systems. Japan was sharply criticized in the last months of 1999 for failing to respond quickly to the Y2K problem. Few of the observers had the courtesy to backtrack after 1 January 2000 to point out that Japan's less frenetic response was actually better suited to the circumstances.

The e-commerce story, however, is far from all doom and gloom. As in most nations, including those experiencing the embarrassing collapse of the over-touted dot.coms, the foremost advances in Japan have been in the business to business (B2B) sector. The structure of Japanese business is, in fact, extremely well-suited to B2B electronic mediation. The highly formalized relationship between suppliers and manufacturers allowed for easy adaption – led by the implementation of e-commerce models by larger firms and their imposition on suppliers –

of digital ordering, monitoring of inventory and delivery, and financial and supply management. Many companies across Japan have computerized their operations, right down to a significant number of the small, neighbourhood production units which underpin a surprising portion of Japan's industrial production.

Japan's commercial structure, especially the top-heavy ties between larger corporations and their smaller suppliers, ensured that elements of the B2B implementation in the country were stronger and deeper than in other nations. Auto manufacturers like Toyota, Nissan, and Mazda, have large and successful B2B operations. Electronic ordering between the automakers has had the effect of creating a virtual marketplace in automobile parts, tying producers across the country into extensive production networks. Osaka-based Daikin provides an excellent, albeit small, example of how the Internet is changing business to business relations in Japan. This fully automated, roboticized manufacturer has long prided itself on commercial innovation. For decades, the producer of air conditioners, heating units and other such devices have built solid, long-term relationships with suppliers. Most of the firms supplying parts to Daikin were Japanese and adhered to the manufacturer's expected delivery performance and high quality standards. The advent of the Internet, which both reflected and encouraged commercial globalization, has rendered the old system untenable. Competitors found cheaper and still high-quality suppliers in other countries and thereby undercut Daikin's market.

The firm believed that it had no choice but to change and to capitalize on the potential of the Internet. At the same time, Daikin refused to compromise on its standards for delivery and product quality; reliability and performance were the hallmarks of its products and justified the higher prices that the company charged. They were not prepared to adopt the free market auction system used by an increasing number of manufacturers. The company merged its old way of doing business with the potential of the Internet. They 'proved' up potential suppliers, conducting factory inspections and ordering sample deliveries, covering most of the costs for the target firms during the process. When and if they were satisfied, the new firm was placed on the list of eligible suppliers.

Daikin now makes most of its major purchases of supplies and components by way of Internet auctions. When need arises for a particular component, the firm sends a notification to all of the acceptable suppliers of the part or product, many of which are now based outside Japan. An auction is fixed for a particular date and time. The auctions do not run for very long, and participating firms can bid electronically, all the while monitoring the progress of the competition. At the appropriate time, the auction is formally closed and the winning bidder identified. Through this system – a variant of the internet auction process which has become commonplace around the world – the company is able to ensure itself of competitive prices while retaining control over quality and maintaining assurances of prompt and proper delivery

In many ways, the Japanese B2B roll-out resembles that of other industrial nations. In size and importance, B2B e-commerce outranks the more highly publicized business to consumer trade by a significant margin. There are electronic job boards and employment services, many of them targeted at hard to find technologists (and English teachers). There are Internet based auction services,[52] advertisers, and content providers. Speciality companies have come into existence to match the communicative power of the Internet with the production processes in the country. The launch of E-Zaiko.com in September 2000 is a case in point. E-Zaiko.com, a joint venture between Mitsui and Itochu, provides an electronic marketplace for producers and distributors of manufactured goods (clothes, furniture, and other items), offering to match sellers with excess product and purchasers looking for items to sell.[53]

Beyond Internet-based initiatives, digital business is a truly global force, with companies as diverse as Microsoft and Nokia developing international markets for top-notch high-technology products. Japanese firms have been quick to recognize the world-wide potential of the technology and services of the new economy, as demonstrated by the mass popularity of such critical exports as *anime* (digitally based Japanese animation that is celebrated the world over for its complexity and quality), entertainment devices such as the Sony Walkman and Playstation, digital cameras, computer hardware, industrial robotics, and copying, fax and video-recording machines. Japanese digitally based technologies are famous world-wide – much more so than Japanese digital services. Major Japanese suppliers – Sony, Matsushita, NEC, Hitachi, Toshiba, and the like – have outlets around the world and attract a sizable share of the global market for key consumer technologies.

On the Internet side, however, Japanese firms have had virtually no impact. While expecting foreign companies to localize their websites to meet Japanese needs, Japanese firms have shown very little interest in preparing web-sites and Internet services for non-Japanese markets.

Japanese e-commerce (save for some important initiatives in the B2B sector) is aimed primarily at Japanese consumers. Much the same, it need be said, is true for most Internet business. The overwhelming majority of companies engaged in e-business work within a single language and most focus their efforts on their home country. The assumption, now disproved, that English would dominate the web and, therefore, e-commerce, convinced many international companies to postpone efforts at localization.[54] As reality slowly settles in on the e-commerce enterprise, there is far greater recognition of the need to prepare Japanese web-sites and back-up services in order to attract Japanese consumers. Alibaba.com, based in China and the centre of one of the most important B2B web-sites, launched a Japanese language site in 2002, providing both foreign firms with greater access to Japanese companies and vice versa.[55] Developments such as this foretell a future of greater international commercial interaction facilitated by the Internet.

Not all Internet-based business is of the dot.com variety. Many of the most successful implementations involve companies which have found ways to use the Internet to augment and expand existing operations. In Japan, one of the best examples of this process involves Seven-Eleven, the country's largest retail chain (8,500 stores and growing). The firm has made regular major investments in information technology management systems; its systems are proprietary rather than Internet-based, but the exchange of digital information is the cornerstone of the retailers operations. The technology links Seven-Eleven with its suppliers, is noted for being easy to use (a key factor given the large number of part-time employees), and is readily adaptable. The implementation, based on NEC hardware, Microsoft Windows and satellite delivery, was completed in 1996 at a cost of some $490 million. It immediately thrust Seven-Eleven to the forefront of Japanese e-commerce.

Seven-Eleven's system allowed the company to adapt quickly to changing consumer needs and to manage its complex operations with ease and adaptability. The firm monitors each of its stores continuously and is able to respond to problems and opportunities. By working closely with suppliers, the company can monitor consumer tastes and develop and deliver new products quickly. (The arrangement even monitors the weather, and enables the company to anticipate sales and store needs accordingly.) The supply chain is monitored extensively, enabling the company to track deliveries and to ensure that its stores are stocked properly. Seven-Eleven also generates instant and comprehensive market analysis. The gender and probable age of customers is entered into the cash register, thus helping the company produce a vast array of timely data on purchases and spending habits.

The Seven-Eleven chain's internally-controlled system does not take advantage of the open-access opportunities of the Internet. Competitive suppliers, for example, or delivery services cannot bid for Seven-Eleven's business and thus provide the company with the best possible prices. But the firm remains committed to its proprietary system, and to its stable of loyal and well-connected suppliers and service providers. Lawson, a major competitor, developed an Internet-based procurement and management system, putting Seven-Eleven's model to the test. Seven-Eleven, however, has capitalized on other aspects of the Internet. Seven-Eleven stores are payment and pick-up stations for Internet companies, responding creatively to the nuances of the Japanese market place (consumers like to pay cash, and daily commutes place a Seven-Eleven store on the path of millions of consumers). The firm claims that 75 per cent of Internet shoppers use stores like theirs to collect purchases – and these same individuals often become consumers in the Seven-Eleven store. The company has even moved into the dot.com business, establishing 7dream.com (with seven other companies) and offering a wide variety of consumer products.[56] Seven-Eleven's success has stimulated competitors to enter the e-commerce field and has attracted imitators in other sectors. It has, at the same time, adhered to existing commercial and employee relationships, and has not used the Internet (as, for example, has Wal-Mart) to usher in a new age of competitive pricing and aggressive corporate profit-management.

In contrast, E-based financial services have been slow to emerge in Japan, although there are signs that this is changing (Seven-Eleven, interestingly, launched a branchless bank, IY Bank, with Ito-Yokado, a supermarket chain). Japanese conservatism around banking and investment is something of a barrier to innovation in this area (while protecting investors from the disruptive possibilities of Internet day trading, a much promoted aspect of the dot.com bubble in the USA).[57] The banks moved slowly for a wide variety of reasons. Their operations are highly structured and rigid and few of the financial institutions had made consistent investments in Internet technology. There are signs that the banking institutions have now begun to respond. The opening of Japan Net Bank, half owned by Sakura Bank (with other investors including Sumitomo Bank, Fujistu, Nippon Life Insurance, NTT DoCoMo, Mitsui and Tokyo Electric Power), demonstrated clear establishment support for the concept. The cut-rate service, launched in October 2000, was expected to be only the first in a series of major financial initiatives, with a Sakura and Sony-led bank slated to open in 2001.[58] These systems are tentative compared to the North American sprint toward the Internet, but promise to be the precursors of more dramatic changes in the future. Moreover, Japanese financial institutions have to contend with the rigidity of the risk-averse Japanese consumers. Many Japanese still avoid credit cards and make only occasional use of electronic banking facilities. Convincing these same people to migrate their business to the Internet is likely to be a formidable challenge.

There have been several major and coordinated initiatives to jumpstart the country's e-commerce activities. One significant example, which attempted to marry product innovation with national service delivery, emerged in the country's book industry. [59] Beginning in the middle of 1999, a consortium of Japanese publishers and book sellers have been experimenting nationally with the introduction of an e-book. Again, great hype and worry surrounds the introduction of electronic books, with corporate enthusiasm not yet matched by consumer interest. E-books have failed to take off globally, despite considerable marketing efforts. In Japan, however, the massive and troubled national book industry has been seeking ways of reaching new markets and avoiding being swept aside in the digital stampede. The Japanese, despite coping with an extremely difficult written language to master, are inveterate readers, and have among the highest rates of literacy and book retail sales in the world (a 1 trillion yen a year industry, involving the sale of 1.5 billion books). Add to this the commuting habits of the Japanese workers – long, boring hours spent on buses and subways – and the absence of domestic space for bookshelves, and the country has a superb test market for electronic books.

Supported by the Ministry of International Trade and Investment, a group of almost 150 Japanese firms implemented a 'book on demand' system. Stations in bookstores and convenience stores around the country offer consumers (some of whom have been given the technology free of charge) the opportunity to download e-books onto pocket-book sized readers. E-Book Japan is designed to test the commercial market for electronic books, which they hope will find favour among consumers and the book-buying public. The system would save a great deal of money in distribution, printing costs and eliminate the losses involved with book returns. Japan is not alone in this endeavour, obviously, as the Open eBook group in the USA negotiated industry standards in 1999.

By 2002, the e-book enterprise had shown considerable promise. The nation-wide market exceeded 1 billion yen and was in line to grow to 10 billion yen by 2005. E-Book Initiative Japan estimated that downloads of books would increase by 10 times between 2002 and 2003. The expansion of the country's Internet capabilities, interestingly, contributed to much of the growth. The newer e-books were not simply digital versions of regular books. Matsushita, for example, introduced a powerful inexpensive e-book device in 2003. Other companies, particularly Sharp, allowed books to be downloaded on personal digital assistants.[60] As in the USA and other countries, Japan's ebook initiative has remained largely grounded, with readers still wedded to the traditional print and paper.[61] Much like the m-commerce revolution, however, commuting patterns of urban consumers, coupled with the portability of the e-books, may create a strong and profitable market for the new technology, particularly if convenience store alliances make the e-book downloads readily available.

Japanese firms are beginning to explore other areas as well. CareNet Inc. began operations, offering to link hospitals, other medical centres and doctors through a service designed to coordinate the flow of patients. The company transformed itself from an information business, based on a satellite television service, into a web-based initiative. The Japanese government provided money for a pilot project in Yokohama, a successful venture which convinced both the government and CareNet that the business plan was viable. The government's enthusiasm, incidentally, originated with the realization that more efficient use of doctors, hospital beds and other services would result in substantial savings. CareNet's web-site allows doctors and hospitals to search for medical service or space at a large number of participating facilities, and enables patients to be routed directly to centres able and willing to provide care. Corporate expansion plans were slowed, ironically, by the fact that well over half of the country's hospitals did not have full Internet connections. At the same time, the advent of DoCoMo's wireless Internet service has enabled CareNet to widen its reach through new web technologies.[62]

In companies around the world, the advent of the Internet has revolutionized commercial operations. For many firms, email remains the main application, and serves as the cornerstone for corporate communications. Other organizations have made major investments in e-business, ranging from internet-based ordering systems to Wal-Mart's massive integration of sales, manufacturing, warehousing, personnel and other commercial activities. The development of enterprise integration software, most notably by firms like SAP and Peoplesoft, represents the logical extension of this development. These tightly bound systems facilitate the flow of information and hence decision-making within and between corporations. To many in the e-business field, enterprise integration represents the greatest Internet-based growth opportunity.

Japanese firms have yet to take to enterprise integration and there is, to date, little evidence of an Internet-based revolution in the management of Japanese companies. Traditional management styles remain much in evidence. E-mail is used regularly, but much less often than in North American and European firms and typically for informal work processes. While computers now appear on most professional employees' desks, the companies have yet to explore the full potential of internet-based re-engineering. It is likely that the wave of staff reductions which hit Japan in 2000–02 sparked greater investigation of e-based corporate operations, particularly in the firms headed by westerners. For the most part, however, the Japanese commitment to employee job security, still a significant feature of national economic life, means that companies are not looking for the work and employee-reducing 'benefits' of e-business.

To a degree that is not widely discussed, enterprise integration software is also surprisingly culture-specific. SAP, for example, is a German firm and the decision-making models and work processes built into the massive software program reflect a specific set of cultural values and assumptions. Likewise, Peoplesoft (which faced a formidable challenge localizing its human resource management software to service companies in different countries[63]), embodies specific American values and ideas about the management of work. Neither system was developed specifically with the Japanese market in mind and hence is not an easy fit with the complex, traditional, hierarchical and consensus-based decision-making processes of the Japanese firm.

The current transformation of Japanese business, driven by market pressures, commercial downsizing, and government attempts to encourage business reform, may well encourage greater use of Internet-based business tools, but the full implementation is likely to be some time in the future. Japanese firms have not shared the enthusiasm of North American companies for managerial efficiencies and cost-cutting through the introduction of e-commerce technologies. The long-standing 'compact' between big business and government and between corporations and employees have slowed staff reductions and delayed corporate commitments to major restructuring. As a consequence, Japanese companies have been much less likely to invest substantial amounts of money in technological solutions, preferring to maintain levels of employment that outside observers consider to be unproductive.

Significant software development centres have emerged, most notably Bit Valley /Tokyo and the City of Kyoto in the Kansai District. Bit Valley is quieter than in its heyday, only a few years ago, but more methodical development of Internet products and content continues apace. Kyoto, renowned as a cultural and artistic centre, has developed a formidable software industry, drawing on the commercial impulse of nearby Osaska, the artistic talent in the city, and significant government and private sector investment in the new economy. Sapporo, building on a long history of personal computer development, a cluster of entrepreneurial activity around Hokkaido University and consistent government support, has emerged as a surprisingly active IT/software centre. In addition, several government-funded university and university-private sector projects have been undertaken around the country, although most of the effort continues to focus on the technology of the Internet and digital revolution, and much less on the commercial applications of these new tools.

Hope springs eternal in the world of e-business, even in the face of corporate failures and declining investment. Japanese firms, well-aware of the clogged highways that dominate the major cities, are exploring ways to link automobiles, the Internet and Geographical Positioning Systems to provide a high-tech highway navigation system for the country.[64] Few companies and few sectors, beyond pornography and gambling, have yet to make significant profits, but the enthusiasm lives on. Within Japan, there is reasonable optimism that deregulation and enhanced competition in the Internet business will bring more consumers on line and lead to a surge in e-business. There is also a sense that aspects of Internet-based commerce are particularly well-suited to the Japanese. In recent years, Japanese consumers have become more concerned about price than in the past, while retaining their strong interest in product selection, quality and service. On the cost front, the Internet provides an extraordinary ability to comparison shop, to check dozens of competitors and to secure the best possible deal.[65] To this point, the opportunity to comparison shop brought out consumers by the tens of thousands to the major commercial areas, like Shinjuku and Akihabara in Tokyo. E-commerce performs much the same function, without the hassle of train travel and the need to wade through massive crowds.

Japan has done more with computerization in business than most countries. In many fields, including robotics, consumer product innovation, and B2B initiatives, it is a world-leader. Japan's greatest accomplishments in digital business lie behind the scenes, in robotics-based faculties, sophisticated internal and business to business connections, and creative product development. Much of its public face – the use of computers for retail purposes, Internet banking, and the like – seems slow-moving by international standards. The country lagged behind in highly touted e-commerce initiatives and seemed destined for a time to be relegated to the economic backwoods in terms of digital retailing. Forecast growth in B2B and B2C e-commerce (Table 4.6) suggest that Japan is likely to retain a leadership role in Asia (with growing competition from South Korea) and will continue to be an important global market. Additionally, the anticipated continued expansion of m-commerce revenues and markets (Table 4.7) provides Japan with growth opportunities that are shared with few nations, with South Korea being a notable exception. Commercial products spurred by the IT revolution do not figure prominently in Japan's international trade, a trend which worries the national government.[66]

Of course, that Japan was late and comparatively unenthusiastic in joining the dot.com revolution meant, as an unanticipated consequence, that the country's investors and companies did not suffer unduly from the bursting of the dot.com bubble. Bit Valley took a nasty tumble, but the temporary success of most of these ventures was not pivotal for they had attracted comparatively little investment capital. In America, the crash in stock prices wiped out hundreds of thousands of investors, principally the late adapters who opted into the dot.com stocks at their 2000 peak. The simultaneous crash of hundreds of Internet and technology firms caused considerable financial distress across America (although it was not followed by the vitriolic and overly critical analysis that followed the similarly dramatic

Table 4.6   **E-commerce revenues and estimated revenues in Japan, B2C and B2B, 2000–2004 ($US bn)**

Year	B2C	B2B
2000	2.2	25.2
2001	5.9	47.4
2002	10.5	76.9
2003	17.3	115.6
2004	24.6	180.2

Source: *eMarketer*, 2000.

Table 4.7   **M-commerce and B2C online trade in Japan**

Year	M-commerce % of total market	Size of B2C market
1999	0	
2000	7.2	Y824 billion
2001	10.8	
2002 (est.)	13.7	
2003 (est.)	15.5	
2004 (est.)	16.2	
2005 (est.)	18.4	Y13.1 trillion

Source: Japan Inc., April 2001.

bursting of Japan's economic bubble; American commerce, it seems, is not very good at self-criticism). Governments and major corporations had, of course, made their investments in Internet projects, many of them as yet unrealized. But here, too, Japan's delay in joining the Internet craze limited the country's liability. There was commercial debris to be cleaned up in the aftermath of the dot.com crash, but Japan found itself facing fewer crises and a smaller number of business failures than in most other industrial nations. There is, it seems, occasional value in being slow off the mark.

Internet technology is based on a simple premise: that information continues to flow around barriers and blockages in search of its intended destination. Packets of data leave the sending computer and, flowing along fibre optic or copper wire, pass through routers and other technological devices. If a line is down or if traffic has backed up the flow of information, the data packet opts for another route, seeking the fastest (not the shortest) possible route. Japan's e-commerce world encountered numerous barriers, in the form of government regulations, the complexity of the telecommunications system in the country, high usage fees and the difficulties maintaining desk top computers in small Japanese homes. The Internet and, by extension, the e-commerce world hates these barriers and seeks ways around them. For most countries, the answer lay in the improvement of data links, particularly through improvements in delivery systems over telephone lines and cable television systems and also through the domestic use of fibre optic cabling. But not, at least yet, for Japan.

Japan had a solution, one which launched the country from the middle of the e-commerce pack and into the forefront of a new form of Internet communication, mobile Internet and mobile commerce (or m-commerce). In a stunning and strategic leap, Japan overcame the limitations of its technological and e-commerce infrastructure and developed new devices and technologies which responded brilliantly to the needs and opportunities in Japan. With both e-commerce and, as the last chapter documented, m-commerce, the development of the Internet reflects national characteristics and needs as much as it does the capability of the technology. When commercial innovation, product development, and technological sophistication match up, as they did with Japanese companies and robotics and as they did again with the *keitai*, the world witnessed a formidable merger of Japanese culture, social attributes and technology.

5
The Digital Face of Japan: National Dimensions of the Internet Revolution

For most Internet users, Japan's digital face is a muddled mess. Without the proper character software downloaded onto the computer, Japanese web-sites appear on the screen as chaotic, illiterate gibberish. And, for most non-Japanese, even when they can receive the information in Japanese, they can make little sense of it. Only a tiny number of non-Japanese people can read the language well enough to make sense of a Japanese website. And despite the heavy promotion of translation software, there are still no widely accessible applications which make the content and services on Japanese web-sites available for general use. For some Japanese sites, there is an escape: many official web-pages have an 'English' button and a minuscule portion offer one or more other languages. The vast majority of e-visitors, having wandered through digital space to Japan, abandon all hope of making sense of the meaningless computer images and scurry back to linguistic safety. As a consequence, Japan's digital face is largely exclusive to the Japanese, with only glimpses provided to non-Japanese readers.

Japan's Internet presence provides an important window into the struggle to retain cultural integrity in the information age. The Internet and related information technologies defy borders and, as a consequence, government regulation. The government of Japan could, and did, slow the advance of the Internet by way of excessive regulation and underinvestment in infrastructure. But as the remarkable success of DoCoMo demonstrates, the Internet despises barriers and seeks ways around digital walls and other technological blockages. Although expense and inconvenience slowed public acceptance of the Internet in Japan, keeping usage rates far below other major industrial nations, citizens have increasingly found their way on line. In the process, they have produced – and demanded – a Japanese web presence. And as numbers of users have grown, government departments, companies, special interest groups, charities, cultural organizations and individuals have capitalized on the Internet as a new, fast and comparatively inexpensive way of reaching large and diverse audiences. From government promotional material through to corporate enticements and new techniques for the preservation of culture, the Internet has arrived as a major means of communication.

The portrait of a people or society on the Internet is fast becoming one of the primary means in which foreigners and nationals come to understand a country. 'Portrait' does not mean that there is a singular vision emerging from the worldwide web, but rather a chaotic mosaic – part propaganda, part commerce, part popular culture and part society in action – which collectively represents the values, aspirations and expressions of a nation. Where in the past scholars looked to central constructions of national identity – through the school system, in major works of literature, museums, cultural celebrations and the like – the Internet now provides a more democratic and comprehensive insight into the nuances and complexities of a particular society. The image that emerges of Japan, trapped in the gigabytes of web-based material, is confusing, unfocused, and contradictory – much like contemporary Japan itself.

The Internet, however, provides unique insights into how a nation constructs itself for internal and external consumption, with the acts of construction coming from people and organizations across the social, economic and political spectrums in the country. A nation's Internet presence is dominated by those web-sites designed primarily for people within the country. Through bulletin boards, chat rooms, e-commerce sites, government offerings, and personal contribution one finds – voyeuristically – a country speaking to itself. These sites are not intended for international consumption, but the technology of the Internet is such that there are generally few barriers (beyond language and the ability to display Japanese characters on one's computer screen) to foreign consideration of the material. Ironically, of course, this reality is quite different than that expected from the Internet in the late 1990s. The assumption was that the Internet would be a major force of globalization and westernization. American English dominated the Internet and e-commerce, and the digital pipelines seemed destined to bring the whole world into the USA's virtual orbit. These expectations have not been realized. English, as of 2002, is still the most common language on the Internet, but it no longer accounts for over half the web-sites. Chinese is expected to dominate the web by

2005 to 2007. Nation-specific web environments are developing around the world, led by Japan and South Korea, but also including countries in South America, Northern Europe and other regions.

The multi-faceted construction of a 'virtual Japan' on the Internet holds many fascinating elements. First, it remains largely inaccessible to people who cannot read Japanese. As a result, it is primarily designed for internal consumption and rarely seeks to inform, educate or transform broader world opinion. There is, by volume, a considerable amount of information about Japan in other languages, most of it provided by the national government, non-Japanese academic organizations, and various non-governmental organizations and companies. (The two best guides to Japanese information on the Internet are J-Guide at http://jguide. stanford.edu/ and Asian Studies WWW Virtual Library at http://coombs. anu.edu.au/WWWVL-AsianStudies.html#Region.) The material is strongly influenced by corporations, through e-commerce initiatives and web-advertising, and is being transformed by the advent of the *keitai*, which is more service-rich than content-strong.

In a related development, the Internet is serving to link expatriate communities as well. Migrants use the Internet to remain connected to their homeland and families. They use the web to follow news from their former home, order food and other goods, and spend hours maintaining contact with friends and family members via email. For those countries with large immigrant populations, the Internet has the potential to slow the newcomers' integration into the society and to maintain bonds with their homelands. A Japanese example provides an interesting illustration of this process in action. Kansai Gaidai University near Osaka has the country's largest international exchange student programme at the university level. The institution provides excellent computer facilities for its students. They, in turn, spend a remarkable amount of time on the Internet, mostly sending emails (and using web-phones) to connect with boyfriends or girlfriends, family members, friends and their home institutions. The time spent on the Internet, all of which reinforces contacts with their homelands, takes away from the opportunity to speak and learn Japanese. As a consequence, students become less proficient in Japanese and spend less time experiencing Japan. The same, it must be said, has become the norm for international students studying in other countries.

In the early years of the Internet, with excitement generating enormous interest, Japanese firms, agencies and individuals rushed into cyberspace. In keeping with the country's long-standing interest in how it is perceived internationally, it is not surprising that many launched English-language web-sites at the same time or shortly after they established a Japanese presence. With Internet access growing rapidly – from two commercial Internet Service Providers in 1993 to over 40 two years later – the number of web-sites grew exponentially.[1] From little more than a standing start five years earlier, by February 1998, Japan was host to over 1 million web-sites, the second highest number in the world and an increase of over 50 per cent from the previous year. (The USA was, not surprisingly, far ahead with over 16 million web-sites. The American military alone had as many web-sites as Japan.) Efforts in this regard benefitted significantly from the availability of a Japanese language version of Microsoft's Windows 95, which included a Japanese language web brower and which brought the Internet into more regular use across the country.[2] However, Japan's per capita ranking in web-sites was considerably lower than the leading nation (Finland, with one internet connection per eleven people), at one web site for every 107 people.

Japan's Internet presence is complicated by the fact that there has been an explosion of *keitai*-based web-sites, which are not accessible over the regular Internet (the reverse is also true). This means, in effect, that Japan has two parallel Internet services, one designed to serve a narrow and technologically defined group of users and one available on a global basis. At present, the *keitai* sites are overwhelmingly corporate in nature, designed to provide services or content to consumers. The continued development of person-to-person contact via the mobile Internet promises to change this orientation as well.

Early usage of the Internet revealed that Japanese surfers followed much the same pattern as in other countries. One 1997 report indicated that three quarters of all subscribers used the web for their hobbies – the largest single usage category for the Japanese web. Close to half claimed to use the web to find product information and over one third to collect business-related data. Almost 1/3 of those surveyed indicated that they used the web to collect 'sex-related' information. One survey of female use of the Internet for shopping purposes revealed that the primary attraction was local food products, followed by books and music. J-List, one of the first firms to enjoy a measure of commercial success on the Internet, offered a wide array of music, books, and personal items. The variety in the catalogue, however, disguised the fact that almost two-thirds of all sales related to what are euphemistically described as 'adult products'.[3]

The Japanese government, as discussed earlier, was a late comer to the Internet and is only beginning to offer public services over the Internet.

Japanese companies are not much further advanced. While there are some successful commercial web-sites, particularly in the retail 'and technology sector, e-commerce has not set out extensive roots in Japan. The most impressive sites tend to be Japanese variants of the leading global firms – Amazon.com, Yahoo! Japan, and the like – and not home-grown products. Japanese companies provide excellent, even artistic, e-pamphlets and the Internet is undoubtedly important in assisting consumers with major purchases. Unlike their North American counterparts, however, which are using the Internet to build customer loyalty, to maintain contact with suppliers and to present an active and dynamic image to the public (including potential employees), the corporate web in Japan is solid, basic and comparatively unexciting.

Major cultural organizations, including museums, art galleries, libraries and archives, quickly understood the web's potential. Japan is, for all of its high technology global face, an intensely historical and cultural society, with a strong passion for the arts and Japanese culture. It is not surprising, therefore, that organizational imperatives (including the need for outreach activities) and public interest have coincided to result in a comprehensive and often excellent cultural presence on the Internet. Again, however, the Japanese sites tend to lack the technological and conceptual creativity of some western organizations, with their interactive museums and dynamic web-sites. Japanese cultural Internet sites tend more toward electronic catalogues – screen after screen of cultural material and information – and include few interactive, 3-D or other innovative presentations of culture. Japanese users of the Internet would have no difficulty locating vast quantities of sophisticated cultural information and countless examples of Japanese cultural activities.

It is in the realm of popular culture, however, that Japan's web presence is most dramatic. Young people, active in computer and interactive gaming, web-design, new technologies, and modern forms of cultural expression, have found the web very much to their liking. The Japanese worldwide web offers numerous pop culture sites, some well-organized and highly structured (most of these academic in design and purpose) but most personal, cluttered and untidy. These expressions of popular culture range widely, from dozens of sites devoted to *anime* (Japanese animation), music, literature, youth events, clothing, body piercing, and the like. Sites pressing specific causes – women's rights, environmental concerns, anti-Americanism (or, more specifically, anti-US military), gay rights, extreme left and right wing politics, and many others – have proliferated, as has the use of the Internet to support charitable and public advocacy work. People with a cause, or people seeking a cause, in Japan will not suffer for want of web material to browse. Japan's aggressively free publishing and political environment has, not surprisingly, been replicated on the Internet.

Although Japan has established a formidable reputation for digital and graphic design in other fields – *anime* being the best known example – creativity and artistic accomplishment are in short supply on the Japanese Internet. The Japanese Internet leans heavily toward the electronic pamphlet approach, although foreign owned or associated Japanese companies (like Yahoo! and Amazon) have adopted familiar designs for their Japanese web sites. There is an irony in this. Japan is a technology-obsessed country, with a penchant for digital wizardry and electronic creativity. Despite the enormous respect accorded the nation's artisans, there are few signs of these skills crossing over into the Internet world. On balance, the Japanese Internet is surprisingly bland and uninspiring, and lacks the dynamism and innovation that the country has lavished on its electronic consumer goods.

The flatness of the Japanese content is not offset by sparkling creativity on the part of websites aimed at the nation's market but prepared by non-Japanese designers. Foreign organizations and companies have a much smaller and less dramatic web-presence in Japan. Language and the cost of translation is a major barrier to web-site design in Japanese; many foreign firms and organizations appear to assume that Japanese interested in non-Japanese products, services, and subjects will read English-language web-sites. (As one illustration of this process, fewer than 2 per cent of Canadian firms intending to sell products into the Japanese market had so much as a word of Japanese on their web-sites.) Moreover, as e-commerce firms discovered to their dismay, providing web textual material in Japanese (or any other language than that of the company itself) means that consumers expected to be served in Japanese (or their language), thus adding to the cost and complexity of offering Internet-based sales. Major government and corporate sites – embassies, leading trade agencies, some foreign Chambers of Commerce – provide multi-lingual web-sites aimed at the Japanese market. In general, however, foreign governments make few efforts to make their material available to the Japanese (and nothing at all like the effort the Japanese government makes to provide core government material to English-speakers, in particular). As a consequence of these developments, the Japanese language Internet – the portion of the worldwide web visited by the vast majority of Japanese users – has only a small foreign presence.

Japan has realized that the Internet serves as a powerful tool for informing the rest of the world about the country and its developments. Beyond the tourism sector, few countries have made as determined an effort as Japan to provide vast quantities of information about the nation, its companies, and its plans to non-nationals. Other countries do this by default; by producing government documents in English, for example, nations like Canada, Australia, New Zealand, the United States, and the United Kingdom provide English-speakers the world over with access to myriad government reports and other information sources. Only rarely, however, do any of these countries provide much more than a cursory amount of information in languages other than English (save for Canada's French language presence). Japan realizes that the rest of the world suffers from a shortage of information about the country, and has moved quickly onto the Internet to provide useful, up-to-date data to users around the world (and particularly to those who speak English). This material, emanating from official and unofficial sources, Japanese and non-Japanese producers, and often available in languages other than Japanese (typically English), provides foreigners with useful insights into contemporary Japan, the IT-inspired cyber communities in the country, and into Japanese society in general.

It is important not to overstate the degree of linguistic expansion and flexibility. Universities, many with active international programmes, typically provide minimal non-Japanese language material. Municipal and prefectural governments, similarly, offer only the most basic information in English and virtually nothing in other languages. Japanese-based companies, except those seeking foreign employees, likewise provide only a few screens of non-Japanese text, if that. Moreover (and this is a very broad generalization), Japanese web-designers pay about as much attention to the grammatical quality of their English-language material as do other Japanese publishers. Little concerted effort has been made to ensure that the content is either readily accessible or in a format that non-Japanese speakers will find attractive and pleasing. (The same, it must be added, is true of many of the very small number of non-Japan based web-sites which endeavour to provide Japanese language content.)

One of the best ways for the foreigner to approach Japan via the Internet is through expat sites based in Japan. The expat community in the country is, by international standards, not terribly large. It tends to be largely transient (particularly the language teachers), cut-off by language and choice from the bulk of Japanese society, professional and technologically proficient, and youthful. Not surprisingly, this diverse group of foreigners quickly seized upon the Internet as a means of establishing a community, or a network of communities, in a country where they were clearly labeled outsiders. Expat maintained sites provide everything from overviews of Japanese society to 'survival guides' for new arrivals. They offer updated lists of activities, including those targeted specifically at the expat community, job-boards, discussion groups, and information on consular services in the country. Individual expats, however, are often well-connected to one of Japan's many sub-cultures, particularly in the popular culture and high technology fields. As a result, their websites often provide first-class introductions to specific aspects of Japanese life, typically with the insights of a participant and an afficionado. Moreover, the expat websites are often written with newcomers and outsiders in mind, and take the time to introduce the web-surfers to the underlying cultural assumptions and norms that support Japanese social practices.

Similarly, the tourism-related Internet sites are designed to ease the transition of newcomers into the country. The Japanese-produced sites are, for the most part, very basic and lack the creativity and content of many international tourism sites. Few, for example, offer booking and credit card payment services; the vast majority are designed to offer fundamental information. There are useful hotel guides, tourist information sites, and event schedules. But Japan's foreign tourism industry is, by international standards, quite small, a victim of the high value of the yen (and the consequent high cost of tourist travel) and the small number of non-Japanese tourists who venture into the country on other than organized trips. Foreign travellers's guides to Japan, therefore, are typically richer in content and useful information than the more mechanistic Japanese-produced material. The foreign tourist sites often provide insightful cultural and historical information, and provide more comprehensive social advice to incoming visitors.

Most non-Japanese people using the Internet to explore Japan will find themselves using non-Japanese sites. There are a large number of academic and cultural sites (the latter largely maintained by museums and galleries) which provide a great deal of information about Japan. (There are similar Japanese sites, but most of them are available only in Japanese.) The academic web sites contain a great deal of sophisticated information, including archival documentation, statistical material, links to Japanese government sites, and searchable data bases on web sites, publications and other information on Japan. While these collections offer the benefits of volume, accessibility for non-Japanese speakers, and ease of use, they present a vision of Japan derived from non-Japanese eyes. As a consequence, there is an imbalance in the general material in favour of selected historical periods (World War II), artistic developments, contemporary Japanese popular culture (particularly those elements which have had an impact outside Japan), and business/economic matters. The latter subject, in fact, dominates the foreign language material and reflects the general international preoccupation with the strength and weaknesses of the Japanese economy and with the difficulties involved in breaking into the Japanese market.

As the Internet grows in educational importance, it is likely that these non-Japanese web sites will remain a primary source of information about the country for students and consumers around the world. There is no comparable Japanese-produced web presence available to off-set the foreign, largely American, coverage of Japanese topics, with all of the narrowness of view and imbedded assumptions contained therein. The Japanese government has devoted a surprising amount of effort to placing English language versions of major documents on English-language web-sites, thus providing a useful and often comprehensive source of information about the country to foreigners. But this is not a terribly effective counterpart to the dominance of non-Japanese web sites about the country and perpetuates a deeply entrenched tradition of non-Japanese learning about Japan through the lenses of foreign observers rather than from Japanese sources.

One is struck, finally, by the manner in which Japanese companies with substantial operations outside the country generally play down the Japanese connection. Major corporations – Sony, Toyota, Nissan, NEC, Hitachi, and the like – maintain a formidable international web presence. Most often, however, they develop country-specific web-pages, designed to respond to the needs of local consumers. There is virtually no systematic mention of Japan in these materials, and no effort made to build a 'Japan brand' through the identification of high-end corporate activity with the country of origin. This is, importantly, in contrast to American, British, German and Scandinavian firms, which routinely highlight national characteristics and alliances on their international web sites. Of course, the Japanese connection to these international corporations is well-known; no intent to deceive is built into the lack of Japanese content. Instead, it is clear that Japanese firms place the highest value on responding to the national interests of consumers, rather than showcasing national accomplishments.

Non-Japanese speakers seeking to explore the country in cyberspace will find that they spend much of their time on foreign web sites, written by non-Japanese people, and on web sites prepared for and by the expat community in Japan (principally Tokyo). Japan has not yet made a concerted effort to present itself to the world via the Internet and has, as it did in the past, been ready to allow foreigners to interpret the country to the world. New technologies – particularly instant translation devices (the translations are less than perfect and sometimes very funny, but they are becoming more workable) – will eventually permit non-Japanese speakers to surf through Japanese language web sites. As they work through the material, they will undoubtedly be impressed by the difference between the two – Japanese and non-Japanese – Internet versions of Japan.

The Internet gap between non-Japanese information on Japan and Japanese language information on Japan presents a formidable challenge for business people attempting to use the world wide web for commercial research. Analysts working in other countries can typically draw on huge volumes of corporate, fiscal and government data and can easily stay abreast of crucial developments. Doing the same in Japan is difficult. The government translates data selectively, and thereby controls much of the non-Japanese public understanding of national plans and activities. (The same was true in the past with printed documents; the general shortage of foreigners able to read and write Japanese have long enabled the Japanese government to influence the flow of information about the country.) Similarly, most of the news and corporate information on Japan available on the Internet is in Japanese, providing ready access in-country and very little internationally. The situation is not all bad. Public services, by several of the leading foreign embassies, and English-language pay sites provide determined business and corporate analysts with a reasonable amount of information – but any serious researcher will inevitably have to pay for private searches and translations for more specialized data. Given the poor track record of most foreign companies and governments in coping with the Japanese language barrier, it is likely that the Japanese language Internet will remain an only partially permeable barrier for some time.[4]

The image persists that Japan lags behind other nations in the Internet 'race', even though the development of mobile Internet capabilities has established the country as the most dominant in the world in this critical area. The 3-G enthusiasm which swept the globe in 2001, resulting in dramatic bidding wars for national 3-G licenses, resulted in only a handful of commercially successful implementations of the new technology, with Japan and South Korea considerably ahead of the pack in this area. Recent improvements in the quality, cost and variety of Internet service delivery (including Softbank's aggressive moves into this market) have improved Japan's comparative situation on basic questions of accessibility. Japan, on balance, does not do that poorly. It is second only to the USA in e-commerce, dominates the mobile Internet field, and has experienced rapid penetration of the home-based Internet market. Falling connection prices and increased competition has ensured that Japan's rate of connectivity continues to expand. In the interim – and contrary to the widely held stereotype that has Japan trapped in the digital dark ages – the country is actually reasonably well-connected. Although the number of computers is smaller than other nations (see Table 5.1), the country has experienced a dramatic rise in recent years in the number of people who make regular use of the Internet (see Table 5.2). The Japanese also have a wider variety of means of gaining access to the Internet. Note that the statistics cited below (Table 5.3) reflect patterns of usage before the widespread adoption of the *keitai* (mobile Internet), which has resulted in much greater and more diverse Internet use.

There is, of course, no single indicator of Internet connectivity or usage which summarizes a nation's level of commitment to the Internet world. Japan's web-sites, for example, are more corporate and organizationally based than most other countries. The individualistic

Table 5.1　Computer access (computers per 100 people)

United States	50
Singapore	34
Hong Kong	30
Japan	24

Source: Nikkei Weekly, 1 January 2001.

Table 5.2　Internet users in Japan, 2002–2004

	No. of Users(m)	*% of population*
2000	17.7	16.2
2001	21.6	19.7
2002	25.1	22.9
2003	28.5	25.9
2004	32.0	29.1

Source: *eMarketer*, 2001.

Table 5.3 Japanese access to the internet, 2000*(%)

Desktop PC	75
Notebook PC	30
Mobile Phone	3.8
PDA	2.0
Game Equipment	1.2
TV	0.7
Word Processor	0.4
Others	0.1

Note: Many individuals have access through more than one system. Hence, the total exceeds 100%
Source: Ministry of Public Management, Home Affairs, Posts and Telecommunications (reported by *eMarketer*, 2001)

and small-business ethos which dominates the Internet around the globe is less pronounced in Japan. Japan, for example, has far fewer Internet hosts than most other nations, and fewer than a third per 1,000 people than the six most active Internet countries, (see Table 5.4). Similarly, Japan's domain (.jp) has remained much less popular than the size of the country and its economy would suggest (see Table 5.5). This changed after 2000, due to the expansion of *keitai* use, the more international use of such domain names as 'com' and greater Japanese involvement on the Internet. The country, save via

Table 5.4 Internet hosts, by nation, 2000 (per 1,000 people)

USA	*234.2*
Finland	159.1
Iceland	130.8
Canada	127.2
Norway	116.5
Sweden	106.2
New Zealand	92.6
Netherlands	81.6
Australia	75.0
Denmark	72.5
Switzerland	63.5
Austria	57.6
United Kingdom	52.5
Belgium	39.7
Italy	32.6
Japan	32.5
Germany	31.7

Source: OECD, *Communications Outlook*, 2002.

Table 5.5 Network domains, 1999

Domain name	Number of hosts(m)	% of world total
Commercial (com)	18.8	33.4
Networks (net)	12.4	22.1
Educational (edu)	5.1	9.1
Japan (jp)	2.1	3.7
United Kingdom	1.6	2.8
US Military (mil)	1.6	2.8
United States (us)	1.6	2.8
Germany (de)	1.4	2.5
Canada (ca)	1.3	2.3

Source: Japan 2000: An International Comparison: http:// www.kkc.or.jp/english/ activities/publications/aic2000.pdf

Table 5.6 Internet users as a % of population, 2000 (approx.)

United States	48
Sweden	48
Canada	43
Singapore	42
Australia	37
Britain	33
South Korea	32
Hong Kong	26
Germany	24
Taiwan	22
Japan	21
Italy	18
France	15

Source: 'Falling through the Net', *The Economist*, 21 Sept. 2000.

the *keitai*, has a much smaller percentage of its population making regular use of the Internet (Table 5.6) and only 17 per cent of Japanese have Internet access at home (Table 5.7). The most active Internet countries have well over 40 per cent of the population with the Internet at home and making regular use of the technology.

Mobile Internet: Japan is the world leader in access to and use of the mobile Internet. As outlined earlier, over 48 million Japanese have access to the Internet via their *keitai*. This compares to minuscule numbers in North America and a sizeable but comparatively smaller implementation in Europe. Japan, in fact, accounted for over 80 per cent of the world's users of the mobile Internet in 2000 (see Table 5.8). Japan's greatest rival at present is South Korea, which has attracted

Table 5.7 **People with Internet access at home, 2000 (approx. % of population)**

Norway	49
Singapore	48
United States	46
Sweden	44
Canada	42
Finland	37
Australia	36
Denmark	35
New Zealand	32
Netherlands	29
Britain	29
Switzerland	24
Austria	23
Taiwan	22
South Korea	22
Belgium	20
Germany	19
United Arab Emirates	17
Japan	17
Italy	17
Ireland	17

Source: 'The next revolution', *The Economist,* 22 June 2000.

Table 5.8 Mobile Internet users, 2000 (%)

Japan	82
South Korea	12
Europe	5
USA	1

Source: eMarketer (reporting results from *Europtechnology,* 2000).

large markets for their version of the *keitai*. This technology is extremely attractive to and adaptable for other Asian countries, which have similar population densities, complex and crowded cities, and heavy reliance on public transportation. Japan and Japanese companies are making a concerted effort to extend their technological reach into this and other areas, particularly in Asia, where they face formidable competition from the South Koreans and superb Japan-like opportunities for mobile Internet applications. As of 2000, well before DoCoMo's growth slowed, the country accounted for over 80 per cent of the world's mobile Internet users, a position of preeminence that they have not yet relinquished.

Internet usage (time): The advent of the *keitai*, which charges according to the amount of data downloaded rather than time connected, makes it difficult to assess the extent of Japanese Internet usage in terms of time. Added to this, long work days, work weeks and lengthy commutes cut into the free time available for recreational Internet use. Despite these constraints, compared to other countries, Japanese computer users actually spend a great deal of time on the Internet (see Table 5.9). The *keitai* is changing this situation significantly, as commuters, school children and workers are finding it easy to integrate the use of the mobile Internet into their lives. Focussing on conventional Internet use (i.e. not including *keitai*), a 2001 survey found that Japanese users spent an average of 7 hours and 56 minutes a month on-line. They were significantly behind Hong Kong (9 hours and 46 minutes) and well-behind the Koreans, who spent a total of 16 hours and 17 minutes a month online.[5] Japanese usage patterns followed standard rhythms, with traffic peaking between 8 pm and 11 pm; the nadir, even for the hard-working Japanese, comes at 4 am. The country demonstrates very extensive Internet use throughout the working day.[6]

As the Table 5.9 reveals, Japanese Internet users are among the most active in the world. Even with the constraints noted above, the Japanese log on more often, stay longer, and visit more sites than web surfers in many other industrialized countries. The Internet-obsessed Americans have the lead in many key categories, in part a reflection of the availability of cheap Internet access. Japanese users,

Table 5.9   Internet usage, selected countries, February 2001

	Japan	*UK*	*Sweden*	*Singapore*	*USA*	*Global*
Sessions/Month	18	13	13	14	20	19
Domains Visited	52	38	33	41	45	47
Time/Month (hours)	9:54	6:27	6:07	8:08	11:02	10:17
Time/Session (minutes)	33:38	20:29	28:21	34:40	33:30	32:38
Duration/Page (seconds)	0:34	0:47	0:38	0:40	0:56	0:44
Current Internet Universe*	51.3 m	29m	5.7 m	2.3 m	167 m	454 m

* All members, age 2 and over, with access to the Internet.
Source: Nielsen/NetRatings (www. Epm.netratings.com). Global numbers are for Jan. 2001.

somewhat surprisingly, are ahead of consumers in many nations that are generally deemed to be more Internet-intensive than Japan.

Internet usage (content): In the pre-*keitai* world, Japanese Internet use adhered closely to international norms, with an emphasis on news/information, email shopping (or, more commonly, searching for information relating to potential purchases), hobby/recreational sites, games, and considerable access to pornographic web-sites. The list of the most often-visited sites (Table 5.10) includes standard fare; portals like Yahoo!, news and information sites, auction/e-commerce sites, and material placed on the web by the major information and communications technology companies. The mobile Internet has resulted in a slightly different focus: email, animation, music, news/information, and commercial information. Mobile consumers are more willing to pay for content (largely due to the easy billing arrangements with the telephone companies) than regular Internet users. Regular Internet users are surprisingly active, more so than the standard image would suggest. Japanese users are active web-surfers, with the average user looking at almost 640 unique pages per month (in the USA, the number is 511) and spending 712 minutes per month online (compared to 554 in the USA).[7] Japanese users, and this is hardly surprising, stick largely to Japanese sites and tend to use the key portals and e-commerce sites that have attracted attention in other countries. Yahoo! Japan remains the most popular in the country, and the major Internet and e-commerce players dominate the top ten lists.

A 2002 study by Interscope, based on a survey of Tokyo Internet users provided a more detailed examination of Internet use (Table 5.11). While search engines remained at the head of the list, users reported

Table 5.10   Top 10 web properties in Japan, February 2002

Web property	Unique audience	Time per person/month
Yahoo!	15,740,000	1.33.33
Nifty	9,989,000	0.21.20
MSN	9,372,000	0.24.03
Rakuten	9,027,000	0.26.41
NEC	8,444,000	0.15.32
Microsoft	7,981,000	0.06.22
Sony	7,493,000	0.10.50
NTT Communications	6,396,000	0.09.30
Lycos Network	5,502,000	0.13.08
NTT-X	5,466,000	0.12.40

Source: Nielsen/NetRatings (www.netratings.co.jp).

Table 5.11   Frequently used web-sites, Tokyo, 2002 (survey by Interscope)

Type of www Site	% of users
Search engines	96.0
News and weather	82.6
Maps and directions	81.0
Online shopping malls	70.4
Travel and leisure	67.3
Sports news	64.7
Dining out	64.0
Computer supplies	61.8
Books/periodicals	59.3

Source: Japan Inc., May 2002.

that information sites (weather, maps, news and sports) rated highly, as did e-commerce and consumer sites.

Demographics of usage: Before the introduction of the *keitai*, Japanese Internet use followed the normal pattern of the industrial world: largely male, primarily teenagers and young adults, tilted toward university students, and with considerable adoption among male business people. The mobile Internet turned this pattern on its head. DoCoMo made enormous inroads with young (mostly teenage) girls and, as consequence, young men. The technology then migrated up the demographic profile, becoming more heavily used by young adult men and women and, only later, by the middle aged and seniors. When these two groups are combined, Japan presents one of the most comprehensive user profiles in the world, although the various groups continue to use the Internet for very different purposes. Compared to the United States, Japan has more older users, a higher percentage of women, and significantly more people in the young adult (25–44 years) category. In America, in contrast, young people are more likely to log on than in Japan, as are women aged 35–44.[8] The *ketai* has, of course, altered these statistics, for the system first attracted a market among teenagers and only later shifted to older users.

Prices and costs: Until 2001/2002, Japanese Internet costs were dramatically out of line with those in other western industrial nations. Connection fees, furthermore, seriously understate the total cost of Internet use for Japanese consumers, as most paid per minute charges for local telephone use. The introduction of competition, the advent of DSL/cable connections, and significant efforts to deregulate the industry have resulted in dramatic reductions in total costs. According to some surveys, Japanese Internet costs are now comparable to North

American rates, although the situation varies considerably from city to city and region to region within the country. According to the *Nikkei Weekly,* in January 2001, total internet access costs (phone line charges and ISP) in Japan stood at $67.12 per month. This was more than double the US cost of $30.02 and the Hong Kong rate of $31.78 and much higher than South Korea ($22.45) and Taiwan ($19.57).[9] An OECD study completed in September 2000 indicated (in US$ purchasing power parity that Tokyo Internet Service fees, including phone charges, stood at a competitive $49 per month. New York was much cheaper ($23.76) but Dusseldorf ($50.71), Paris ($59.50), and London ($60.41) were higher.[10] There are many different calculations of comparative Japanese Internet costs, with considerable disparity among them. The best way to understand the Japanese scene is this. Until 2001, Internet costs were, on a global comparative basis, extremely high and served to deter Internet use. Competition, the *ketai* revolution, and the introduction of new technologies (particularly cable Internet) lowered prices dramatically in the major cities, although service, cost and quality vary widely across the country. The more favourable cost structure has attracted more companies and consumers to the Internet and future price declines are anticipated.

The Japanese Internet reflects the country in all its strengths, weaknesses and eccentricities. Starting well behind North America, Japan gradually addressed its technological and regulatory short-comings. As of 2002, the nation's Internet was as fast and cost-effective as most in the industrialized world and usage rates were creeping up toward international standards. Equally important, the government recognized the inherent dangers in ignoring one of the most important technical revolutions of the last 100 years and, belatedly, threw its weight behind the Internet enterprise. There is, however, little national interest (beyond Softbank) in using the Internet to connect with other peoples, markets or constituencies. Japanese web-sites are designed for Japanese users, primarily within the country, and limited effort has been devoted to using the web as a launching pad for the international extension of Japan. This does not hold for the technology of the Internet, however, and the Japanese government has made major investments and foreign aid contributions in order to ensure that Japanese Internet technologies play a major role in South East Asia.

Despite its slow start, Japan now stands at the forefront of the Internet revolution. The content elements of the mobile Internet are not readily exportable. If and when DoCoMo and other Japanese firms develop mobile Internet applications in other countries, only a few of which stand to benefit as much from the system as Japan, it will be the hardware not the websites and e-commerce models which transfer to other nations. I-mode and its competitors is ideally suited to the Japanese market and represents a uniquely Japanese blend of technology, marketing, and consumer response. The system requires minimal technological skill, has both personal and professional applications, and is inexpensive to use. As a consequence, and to a degree that much of the world has failed to recognize to date, the mobile Internet is making major inroads into Japanese society. Japanese firms, again led by DoCoMo, have already moved beyond current m-commerce models and are increasingly looking at machine to machine and person to person applications for i-mode technology. In the process, they are broadening and re-defining what the Internet will ultimately mean in the lives of people the world over.

As the world-wide web continues to evolve, the representation of individual societies and cultures will change dramatically. In the first years of the Internet, cultural nationalists the world over, particularly in Japan, China, and France,[11] worried that the new technology represented only the most recent variant of Anglo-American cultural domination. The world wide web, these nationalists believed, would lead young people, in particular, away from national preoccupations, expand the impact of American popular culture, and erode domestic language skills. France, one of the most culturally paranoid nations, was particularly preoccupied with the potentially destructive impact of the Internet and worried openly about how the new system would undermine the French language. Over the first decade of the Internet, the concerns appeared to be well-founded, as English users dominated, American web-sites proliferated, and other languages languished on the margins of the world wide web. As global connectivity increased – and contrary to the deep concerns expressed in the initial start-up of the Internet – the world wide web assumed a more multi-cultural character. There is increasing evidence that the Internet will not overcome linguistic and cultural barriers. Indeed, to the degree that Japan is representative of broader patterns, the success of the Internet in any particular nation depends on the ability of governments, corporations and the Internet community to adapt the technology and content to meet local language and cultural needs.

Corporations discovered that English-only web-sites did little to help them penetrate international markets. Mirror sites, offered in a variety of languages, proliferated. Governments, anxious to spread news about their services, programmes and initiatives, offered mirror sites as well, typically providing information in the national language(s) and English. Many commercial websites sought to draw in users from around the world, and discovered that providing English-only coverage was a major liability. (As mentioned, the most innovative and commercially successful websites are typically those related to gambling and pornography. Adultcheck.com, the most popular of the pornography portals, provides its services in English, Japanese, German, Chinese, Spanish, French, Italian and Portuguese.) Companies have discovered, as well, that providing services in more than one language adds significantly to the cost of doing business internationally. The global reach and multicultural potential of the Internet has not yet proven to be the profit centre many early promoters expected.

There are significant technological problems with providing Internet sites in Japanese. Modern computers, although significantly advanced over earlier models, can still handle fewer than 20,000 kanji. Estimates suggest that more than five times that number are required for a proper kanji-based system. Moreover, the limited complexity and flexibility of the current system is compromised further by the tendency of operating systems and programmes to compress Japanese, Chinese and Korean characters into a single kanji. As University of Tokyo computer scientist Ken Sakamura observed, 'IT is an instrument for greatly increasing the efficiency and distribution of information. To make that a reality, however, computers must be capable of handling all words used in various regions. Already a BTRON/Superkanji system that can handle over 100,000 words has been developed. Discussions of creating a "cybersociety" in Asia can have practical meaning only when such a kanji-based operating system has been developed. This view is gaining ground not only in Japan, but also in China where work is under way to develop an operating system for kanji characters.'[12]

The rapid expansion of non-English web-sites, including the growth by Chinese, Korean and Japanese language domains, has created a unique problem for the world wide web. Hardly surprisingly, Asian Internet users wish for Asian-language domain names (urls), and wish to use Asian characters and spellings. To this point, the Internet's largely English origins provided for a common linguistic base for domain names the world over. By 2002, Asian language domain names and kanji-based urls were in use in Japan and elsewhere, establishing a further barrier to international access to the Japanese portion of the Internet. As Asian language domain names and kanji-use proliferate, coordinated through the internationalized domain phones service.

And there is ever indication that they will, the basic universality of the world wide web will come into doubt.[13]

For the most part, the multi-lingual Internet remains global in reach and orientation. French language sites aim to reach the widely dispersed Francophone/French-speaking populations in France, Quebec (Canada), Africa, Vietnam and elsewhere. Similarly, Spanish, Chinese and German web-sites are of utility far beyond the boundaries of a specific nation, and address language groups scattered widely across the globe. The development of a truly multi-lingual and multi-national Internet is, likewise, a boon to migrant communities, enabling them to stay in touch with family, friends and community when living thousands of miles away. This will, in turn, facilitate e-commerce activities aimed at ethnic disasporas. Migrants now routinely stay in touch with home country events through Internet newspapers, magazines, radio stations and television programmes. As time passes, this information connectivity will be enhanced to include commercial, social and educational contacts with the country of origin. But – and this is an important consideration – very little of this applies to Japan.

Japan's famed heterogeneity and the limited migration of Japanese outside the country (at least since the 1930s) has restricted the growth of Japanese-speaking communities in other nations. With the major exception of Brazil, where there is a large Japanese-Brazilian population centred in Sao Paulo, there are only small concentrations of Japanese people outside the country. (This was not the case early in the 20th century, when large migrations of Japanese people to Canada and the United States gave Japanese migrants a high profile in those nations. In both Canada and the USA, however, the Japanese immigrants assimilated into the mainstream and only some of the ancestors of the original migrants still speak the language.) While Japanese managers dispatched to overseas plants and offices have clearly capitalized on the Internet and new information technologies to maintain strong contacts with Japan, there are no growing Japanese communities in other countries that stand to benefit from and contribute to the Japanese language worldwide web in the manner in which Chinese, Spanish, German and French speakers around the world are assisting with the linguistic diversification of Internet web-sites.

Two countries, Japan and South Korea, are developing what is substantially a nation-specific Internet presence. In contrast to other linguistic cultures, which are dispersed widely and which how have access to an Internet which reflects geographic, economic, political and cultural diversity, Japan and South Korea are building a worldwide web network which reflects their particular culture. Setting aside the largely academic and cultural sites maintained outside the country by universities, museums, galleries and other such organizations, there is very little Japanese or Korea material – and much, much less Japanese or Korean language material – being added to the Internet in other countries. As the websites change and mature, the Internet will increasingly serve the local population. The sites will reflect internal diversity, giving an opportunity for marginal and disadvantaged groups within Japanese and Korean society to promote their vision of their country to a broader audience, providing advertisers with access to a discrete and largely homogeneous population, and offering governments an excellent opportunity to promote national cultural solidarity. Once feared as a tool of western cultural imperialism and the 'dumbing down' effects of North American popular culture – an influence still to be reckoned with in all countries with unencumbered access to the Internet – the worldwide web is emerging in Japan and South Korea as an informal re-enforcement of national culture.[14]

Linguistic barriers slow communication and international understanding; they also create business opportunities. Japanese firms have invested heavily in computerized and on-line translation services, including text and voice systems. A variety of companies, from Virtual Ex-Pat, which establishes a Japanese 'identity' for companies not actually based in Japan to Export-Japan.com, have emerged to work across languages using the Internet. Export-Japan.com serves an intermediary role, providing foreigners with access to hard to find and unique Japanese products and helping importers finds potential buyers within Japan. As automated translation capabilities improve – and early indications are that developments are proceeding more slowly than anticipated – these firms may find themselves displaced. For now, they play a vital role in bridging language barriers and helping companies find ways in and out of Japan via the Internet.[15]

The services, coverage, and content available on the worldwide web will continue to evolve, but not necessarily in a manner consistent with government intentions. As it unfolds, uncertainty about the intent, impact, and controllability of the Internet will increase. Furthermore, as the long-standing debate about the fate of the Napster music exchange system demonstrates, the Internet is substantially impervious to state-regulation and regular government discipline. (A US court ordered Napster to close early in 2000, due to breaches in copyright, but the technology proved difficult to stop and, even as the struggle with Napster continued, other websites offering similar services opened.) The Internet will develop, nationally and internationally, as a result of inchoate social pressures and market influences, not through central dictates or elite prerogatives. In the infancy of the worldwide web, Japanese officials worried that the English-dominated content would spur the Americanization of the country and cut into Japan's cultural homogeneity. While elements of this pattern can be seen, the general trend has turned out much differently.

Language and culture have proven to be pivotal to the development of the Internet.[16] According to Carrie La Ferle et al, 'currently, very little literature exists that draws a direct relationship between the cultural aspects of a country and the rate of internet adoption. However, borrowing from the literature on new-product diffusion rates, various cultural dimensions associated with a variety of countries have been found to influence the process... . All things being equal, once a product is accepted in a high-context culture such as Japan, it is said to spread rapidly because people in these cultures want to maintain an image that is similar to the group.' After reviewing the international literature on new product adaptation, the authors concluded that 'the fact that Japan is a collectivist culture, with high uncertainty avoidance and large power distance characteristics, helps to explain the slower adoption process of computers and the Internet in Japan. ... Collectivist cultures, such as Japan, often prefer to do things as a group and therefore adoption of the internet should be slower in the initial stages than in a more individualistic culture such as the United States. Risk averse cultures, such as Japan, should also be slower because more time is necessary to assess potential pitfalls. Masculine and hierarchically structured societies, such as Japan, place a lot of emphasis on success and protecting the image of one's title and/or family name (i.e. Saving face) an thus may avoid the internet until its usefulness and the process of how it works can be clearly ascertained'.[17] While this analysis has certain attractions, it ignores the completely contrary adoption rates for the mobile internet and other, more well-suited technologies, in Japan. Practicality, as well, as culture serves as major determinant of internet use and innovation.

Language also played a critical role in the development of the Japanese Internet. In Japan's case, this has meant that a Japanese language web has emerged, created almost entirely within Japan and focused almost exclusively on the interests of people within the country. There is considerable Japanese use of non-Japanese web-sites – many of the best entertainment and technology sites are in English – the reverse is not true. Only a small number of non-Japanese speakers make regular use of Japanese language material on the Internet. Those who do access Japanese material do so through a large number of foreign-mediated sites (largely academic and professional) and via the Japanese government's extensive English language web presence. Non-Japanese speakers using the web to learn more about the country typically find themselves drawn to high-brow cultural sites, content heavy government sites, and expat bulletin boards and information centres.

By relying on these standard sources and being unable to capitalize on the large and growing volume of Japanese language material on the worldwide web, non-Japanese speakers encounter a very different cyber-Japan than that available to the Japanese. They do not see the diversity and techno-chaos that is the Japanese language web world and therefore do not experience the degree to which the Internet permits insiders to explore the full complexity of contemporary and historical Japan. The small entrees made by foreign language companies and web-sites scarcely scratch the surface of the Japanese Internet;[18] similarly, the small number of Japanese sites providing material in translation offer only a partial glimpse into Japan's web presence. Overwhelmingly, Japan's face on the Internet is in Japanese, by Japanese and for Japanese.

Japan's Internet presence tests one of the founding assumptions of the worldwide web – that it would speed globalization, overwhelm local cultures with international popular culture, and threaten the coherence and sustainability of distinct societies. As with e-commerce, actual experience has deviated substantially from initial expectations. And while the divide between Japan and the rest of the world scarcely matches that of the nation before the arrival of Admiral Perry's 'Black Ships' in the mid-19th century, language and culture have once again provided Japan with a formidable barrier to the rest of the cyber-world. Japan has capitalized on technological advances but appears, as in so many other aspects of modern Japanese history, determined to preserve national cultural integrity through the development of a Japanese language web presence which reflects and stimulates contemporary Japanese society.

Japan's Internet is, indeed, Japan's. The country is home to one of the world's most important nation-specific Internets, with language serving as a formidable firewall between Japan and the rest of the world. Japan's web-presence is designed to address the specific needs of Japanese consumers and citizens and, with rare exceptions, is not designed to either welcome the world or present the nation to citizens in other countries. In many important ways, Japan's Internet presence foreshadows global developments on the web, ones which run counter to the long-standing assumption that the Internet would be a major culturally integrative force. Language and the very nature of the Japanese Internet, particularly the heavy reliance on the *keitai*, reinforce the separatist nature of Japan. It is increasingly evident that, as in Japan, the Internet is serving local, regional and national purposes and may well not be the major globalizing influence that proponents and critics alike believe it to be.

6

Reflections on a Networked Nation: Japan and the Future of the Digital Revolution

Computerization is now commonplace in the western industrialized world. Most of the computers in use are almost completely hidden. They run car engines, manage medical machinery, control electrical and water supply systems, provide the foundation for manufacturing processes, deliver email and otherwise service, support and influence everyday life. In this regard, Japan is not much different from any other industrialized country. Even the determination of the Japanese government to enhance the role of Information Technology within the country bears a striking resemblance to the national agenda in virtually every country on earth. (Even Myanmar, the Albania of the 21st Century, recently proclaimed its commitment to information technology.) The application and impact of IT, however, varies dramatically from nation to nation; and the ability of a country to become truly 'wired' is increasingly seen as a precondition for success in an evolving globalized and technologically driven economic and social world.

The fascination with technology in recent years has drawn attention away from culture and social values as a critical element in economic performance and national competitiveness. The ethos of high technology has, at its core, the assumption that access to technology, by itself, can propel a nation forward to the next level of prosperity. Laying fibre-optic cable has become the modern variant of the Field of Dreams: install it and they will use it. Putting computers and Internet connections in classrooms will, the optimists say, produce a future generation of computer-capable workers. Introduce competition to lower the price for domestic Internet service and consumers and companies alike will race to the web. Create the technological foundation for interactive video and web-based television and an e-citizenry is just around the corner. Just as certainly, the comparative absence of technology is deemed to be a key indicator of economic decay and marginalization. Failure to build national Internet capacity and the country, it has been argued with forbidding certainty, is doomed to languish. Train only a small number of Internet technicians, likewise, and the country's economy is doomed to second-tier status, at best.

As the digital age progresses, however, the confluence of culture, values, technological sophistication, and national will is increasingly being seen as a crucial component in the evolution of nations. The title of a recent book – *Culture Matters* – captures this sentiment very nicely. More specifically in the context of Japan and the digital revolution, culture matters in terms of the receptiveness, creativity, and long-term adaptability of a country to the economic and social opportunities presented by the Internet and related developments. Technology, by itself, does not generate change. The globe offers countless examples of underutilized technological systems and implementations; the technology 'fix' is a vestige of post-World War II pre-occupation with modernization. The availability of the Internet is one thing; making economically and socially productive use of the Internet is something else again.

Examples abound from around the world. The United Nations is increasingly worried about the 'digital divide', and fears that computerization will widen the gap between rich and poor nations. India, particularly the area around Bangalore, has demonstrated that new technologies, when combined with appropriate educational initiatives, provides regions and countries with opportunities to leap ahead of competitors. Malaysia has tried this as well, tying its economic future to the idea that technology will permit the nation to jump a stage in economic development – from an agrarian economy to a high technology and service economy – within a few short years. Political turmoil and investments of marginal value have derailed the Malaysian vision somewhat. Other nations – Finland being perhaps the best example – invested heavily in infrastructure and training and joined the technological mainstream. Scandinavia is now one of the most extensively wired regions in the world and appears poised to capitalize on the endless opportunities of computerization. Canada, in contrast, has an excellent Internet infrastructure, but appears to lack the entrepreneurial drive, risk capital and educational focus to make the most of the digital revolution. The United States of America continues to lead the world in the commercial applications of computerization, and remains a seemingly inexhaustible source of venture capital (although the opportunities have diminished considerably in the aftermath of the bloodbath of dot.com stocks in 2000); but in America, questions remain about the public acceptance of Internet-delivered services, the reach of the Internet across social, economic and racial lines, and the ability of business to create viable e-commerce business models.

Countless observers have argued that national economic success depends on the ability of a country to invest in computer technologies and infrastructure, develop educational systems which train students for the new realities, and convince businesses to capitalize on the opportunities now available. To the degree that this is so – and the evidence is pronounced that high technology will be critical, if not determinative, to economic development – it becomes essential to explore the national response to computerization and information technology. Putting aside the endless debates about Japanese innovation (or lack thereof) and government responsiveness to the IT challenge (and two decades of resistance to the Internet), a basic question remains. Is Japan well-situated to capitalize on the opportunities of the Information Technology revolution? One answer, offered with a healthy dose of caution, is that Japan is well-placed to tackle the IT transformation. As argued earlier, the country plays a pivotal role in the manufacture and development of consumer, industrial and other IT appliances and technologies. But the reasons for optimism are more deeply rooted in a sense that Japanese society, values and social structures – much more so than in many other countries – are particularly well-suited for the new realities of computerization and information technology. Culture does matter, in the application of new technologies as in other areas, and it follows logically that aspects of Japanese culture will condition and influence the country's ability to capitalize on the potential and avoid the pitfalls of the digital revolution.[1]

Consider, for example, a partial list of examples of how Japanese realities might affect the nation's receptiveness to IT-based transformations:

1. The pattern of accepting government directives (diminished in recent years) will influence the willingness to accept government services delivered via the Internet. Japan is less individualistic than most western nations, values loyalty to the nation above individual desires, and demonstrates greater support for government initiatives than most other industrialized countries. Japan's financial woes since the mid-1990s tested the public's resolve in this regard, and there is considerable suspicion about the ability of politicians

and the civil service to lead the country out of its current difficulties. The early response to the selection of Prime Minister Koizumi, however, suggested that, as in the past, the willingness to follow the government's lead remains a significant element in the national psyche.

The situation is not completely clear, however. The chaos of national politics over the past twenty years has preoccupied politicians and forced radical changes in the civil service. These forces have weakened Japan Inc. at a time of intense international competition and has resulted in risk-averse leaders rising to high public office. While countries as diverse as Finland and New Zealand, Canada and Malaysia, touted the Internet and the new economy as the cornerstone of national economic success in the 21st century, the Japanese government lost the resolve that had created the synergies of public administration and business, leaving the country rudderless at a critical juncture. The Mori government, which ended early in 2001, was the epitome of ineptitude and incompetence, and clung to office only through the historic strength of the LDP and the weakness of the divided opposition. The political culture of Japan has, as was argued earlier, interfered significantly with the evolution of the digital revolution in Japan. The declining status of the national civil service has, consequently, reduced the government's ability to generate support for the new IT initiatives. Scepticism may be difficult to change.

2. The conservatism of the Japanese business community slowed the response to e-commerce and left the country behind American innovations in this area. International dot.com promoters ridiculed Japan's tepid response and mocked the aspirations of Bit Valley in its attempt to replicate the Silicon Valley success. The passage of time revealed, however, that that same conservatism insulated the country from the over-excitement and shoddy investments of the dot-com fiasco. The country lost out on the creativity sweepstakes; its response to the potential of the Internet was imitative and largely unimaginative (until DoCoMo came along). By waiting out the hype of the early e-commerce revolution and the scurry to software development and new business models, however, the country unintentionally avoided the full impact of the high technology meltdown that hit the United States in 2000/2001.

3. The critical importance of commuting and the related daily travel schedules of millions of Japanese workers, particularly in the

Kanto (Tokyo region) and Kansai (Osaka, Kyoto, Kobe) areas shaped and informed the country's reaction to the mobile Internet and e-commerce. In the latter case, the manner in which the company's retail sector has adapted to commuting (major stores attached to train stations and the wide reach of convenience stores) provides Japan with a ready-made distribution system for e-commerce purchases. This system, importantly, allows consumers to pay cash on receipt of the goods, thus overcoming the Japanese long-standing reluctance to use credit cards on-line. The long hours devoted to commuting also created fertile ground for e-books, DoCoMo and m-commerce. E-products and e-services that are well-adapted to commuters' needs have done well in the Japanese marketplace, and a sizeable Internet-based industry is developing rapidly around the juxtaposition of mobile telecommunications and Japanese commuting.

4. The role and position of Japanese women may well prove to be an essential condition to the long-term success of e-commerce, which has been hindered in other countries by the absence of a delivery system which matches with the movements and habits of adults. The percentage of adult women who are home most days is markedly higher from the situation in much of North America and Europe, where they have been entering the paid workforce in larger numbers. This means that home delivery models in Japan, unlike most other nations, can use the full daytime period and not just the critical few hours when families gather before or after work. This, combined with the regular nature of (largely male) commuting, provides e-commerce companies with several ready avenues for matching Internet-based shopping and home delivery.

5. Japan's preoccupation with protecting national language and culture from foreign (especially American) influences is decisively threatened by the availability of the English-dominated Internet, or so the early observers of the Internet believed. This served, in the early years, as a powerful incentive not to invest in information technology, for the fear of cultural globalization, and specifically Americanization, has long concerned Japanese authorities. Early attempts to capitalize on the potential of the Internet – such as the First Generation Computer System Project in the 1980s and 1990s which sought to connect students across national borders – demonstrated the unsuitability of the English-dominated web for Japan.[2] Conversely, the rapid expansion of Japanese

language web-sites and computer services in the second half of the 1990s sparked a rapid growth in Japanese language Internet usage.[3] The intrusive potential of the web could not ultimately be denied, although the resulting developments did not accord with initial fears. Language proved more important and resilient than initially thought, and language-specific Internet networks emerged over time. The current trend toward the creation of a sizable, contemporary Japanese language web presence may actually prove to be a critical means of insulating (and isolating) Japan from the rest of the web-world (which has already begun to shift toward a multiple language universe). This, ironically, is the reverse of what was initially anticipated.

6. The technological creativity of Japanese consumers and businesses helped establish the world's best test market for IT and computer-based innovations. New products are readily adopted by Japanese consumers, and national businesses have responded with a steady wave of innovations. If the country's major companies can capitalize on this significant creative advantage, the opportunity exists to remain a world-leader in consumer electronics and industrial applications. The lead already established in such critical sectors as digital cameras, photocopiers, computer games, navigation systems, and music devices (from cassette players to DVDs to digital music storage systems) could, in time, extend to computer-mediated household appliances, advanced automotive devices,[4] and other such products. It remains to be seen if the pattern of innovation will extend to non-consumer applications, including computer-mediated education, tele-health, and e-government. The initial indications are that a web of cooperative ventures involving entertainment and communications firms, among other combinations, are stretching the market for Internet-based services. Sony and NTT, for example, paired up to use the Sony Playstation to receive video content (games, movies and the like) via the Internet, thus removing the often complex and cumbersome personal computer from the Internet equation. Omron and Matsushita Electric have invested in remote health care systems, as have Sanyo and NEC. These firms and others clearly see the marriage of technology, changing demographics, consumer receptiveness and strong government interest in controlling the cost of health care as a critical market opportunity. The development of G-Book by Toyota, Denso and KDDI provides an in-dashboard

computer for automobile drivers, offering both GPS navigation and variety of web-based services (including traffic reports, weather updates and financial information). The data stored in the G-Book can, through the use of a special card, be transferred to a portable computer or PDA, providing even greater portability and flexibility.[5] There are many other examples, each demonstrating the willingness of major companies to invest millions of yen in an attempt to prove the viability of a new market nice.

7. The remarkable growth and sustained demand for the mobile Internet reflects the degree to which the Internet has to be connected to social, economic and cultural realities in order to be successful. PC-based Internet use in Japan remains comparatively limited, while the *ketai* has become an integral part of national life. This mobile technology – cheap, easy to use, accessible, not place bound and exceptionally well-suited for the commuting lifestyle which dominates Japan – has not been readily transplanted in other countries. Within Japan, however, the *ketai* 'revolution' has brought the Internet into broad use, lessening the 'digital divide' which has plagued many other countries.[6]

These are but a few examples of the manner in which Japanese social and cultural characteristics have influenced the development of the IT revolution. Culture will continue to effect the development of Internet and computer applications in Japan and will ultimately determine the degree to which the country is transformed by the digital revolution. As in all countries, messages and cultural realities are mixed; some reward innovation and technological creativity while others will stifle the imaginative and cost-effective application of computer and Internet-based solutions. On balance, however, Japan appears to be particularly well-suited to exploiting the full potential of the digital revolution.

The relationship between culture and digital transformation may prove to be of fundamental importance to determining the comparative success of nations in the new technological order. It has yet to attract much attention. Speculatively speaking, it appears as though certain countries (the USA, Finland and Sweden) are open to selected consumer uses of the Internet. Others, like India, Thailand and Taiwan, are prepared to place their education systems at the disposal of the IT agenda – thereby taking a huge and underappreciated national gamble on their economic future. Western democratic nations – Canada, the

USA, Australia, New Zealand, Britain and the like – make noises about IT education, but lack the national will to pose high scientific standards on their schools and students. The potential for these nations to fall behind in the technological race is very real. Centralized governments, and authoritarian regimes, have the political capability of requiring public use of e-delivered government services; convincing citizens of more democratic nations to accept similar arrangements will likely prove to be far more difficult. (Much more has been written about the potential 'democratizing' impact of Information Technology and the Internet than on the system's capability for centralizing the control and delivery of government services.) For the last decade, research, government attention, and critical thinking has focused on the nature of the technology and the potential applications of new developments. Far too little time has been devoted to exploring the societal reaction to e-mediated services, products, and information. This study of Japan's response to the IT revolution is predicated on the belief that some countries and societies will be better able – for reasons of political culture, values, social patterns, and approaches to education – to respond to the opportunities presented by the computer revolution. It is not at all clear, as conventional wisdom currently has it, that western industrialized nations are the best situated to become true e-countries.

No one, not even the best analysts of technological and national development, knows how the information revolution will unfold. Technological innovations continue at a remarkable pace, although risk capital may be in short supply in future years as the markets adjust to the false promise of the Internet boom. But there is a gap – potentially enormous – between what technology permits and what society allows. Many great inventions sit idle because consumers found them uninteresting, too expensive, too difficult or too threatening. Add to this all the confusion about standardized technical protocols (the VHS/Beta debate of the 1980s does not hold a candle to the many arguments about international protocols for everything from wireless phones to security systems), uneven distribution of Internet and computing infrastructure, the fads and foibles of national governments, the whims of consumers and deeply seated suspicions about the international flow of personal and corporate information. At the end of the day, all that can be truly known is that things will change.

How Japan will fare in the continuing digital revolution is particularly uncertain. The national government has joined in on the enthusiasm, and is making major strides in regulation, infrastructure development, and the computerization of government services. The rhetoric of change and national innovation matches, in noise and intensity, the promises and exhortations associated with the economic restructuring of the 1960s, 1970s and post-Plaza Accord years. Japanophiles see in the current enthusiasm of government and business signs of old Japan Inc., that determined and formidable blend of government, business, consumers and tax-payers, mobilized behind the common goal of ensuring that Japan remains a major world economic leader. For those who continue to forecast the country's economic downfall, the haphazard maneuvers of the sloppy Moro government are indicative of a country lacking direction and purpose, slavishly following international trends which may not suit Japanese needs or abilities.

Underestimating Japanese adaptability is a long-standing western tradition. The country was hurt, but not demolished, during the recession of the 1990s, and the company's corporations remain among the most technologically innovative in the world. Equally, the wealthy and risk-taking Japanese consumers provide a perfect counterfoil for the digital age, as they are eager to test – and reject when appropriate – new products introduced by Japanese and other producers. Japan has considerable advantages in the race of digital competitiveness, including corporate design and manufacturing creativity, market-savvy retailers, a densely-settled population (which makes the diffusion of technological infrastructure comparatively easy and cost-effective), large pools of local investment capital, ready and well-informed access to the fast-growing Asian market, and a wonderful merging of creative and technological energies in the digital arena. Government involvement, in this scenario, is a valuable but not necessary condition to success in the Internet age. And this, to date, describes the Japanese approach to the development and implementation of information technologies – company and consumer-led, with the government pulled reluctantly along.

There is growing evidence of more community-based efforts to capitalize on the opportunities presented by the Internet. Much of this, as in most countries, consists of local training programmes, activities by small organizations to shift their information on-line, and government, corporate and community initiatives to connect with e-enterprises in the fashioning of a new economy. Within the country, there are a large number of activities designed to make the Internet and its major tools (the world-wide web, email, geographical positioning systems and the like) into regular aspects of Japanese life. Many of these capitalize on the skills, expertise and work habits of retirees and mobilize this considerable resource in the interests of creating an e-Japan. Kyoto's Miako Net established wireless LAN access points to provide free Internet access. SeniorNet Japan (based on an American model) in Fukuoka Prefecture has been using the Internet to promote regional culture, through the identification of tourist destinations, cultural activities, historical events, and related information.[7] In other communities, volunteers and community organizers work at establishing Internet-based companies, hoping that the digital economy can stimulate otherwise declining regional economies. Protest groups, including those agitating against American military bases in Okinawa and against the Nibutani Dam in Hokkaido, have capitalized on the Internet as a means of mobilizing opposition to government (with relatively limited effect to date). [8] New religious movements have been proliferating in Japan and many have capitalized on the communicative potential of the Internet to attract potential converts and to remain in contact with communicants.[9] While these initiatives, and dozens others like them, are broadening the base of Internet activity in Japan, there remain considerable gaps in the mobilization of the society and economy for the digital revolution.

This pattern has resulted in Japan becoming an active, but clearly second-tier, player in the practical applications of the digital revolution. Save for an enormous and underestimated lead in mobile computing and digital devices for home use, Japan lags well behind the United States, Canada, Finland, Sweden, Australia and several other more aggressive computer nations. Perhaps the most serious shortfall is to be found in the business to business field, where American firms have an almost insurmountable lead that provides the United States with a critical national economic advantage. Japanese companies and *keiretsu* are racing to catch up, and major corporations are rapidly dragging their network of parts suppliers and affiliated companies into the digital age. At the consumer-level, NTT's phone charges, scheduled to continue dropping, have placed a strong brake on home Internet use, dampening enthusiasm for everything from e-commerce to computer-based education. Oddly, then, other countries are using Japanese devices – personal and portable computers, personal assistants, television sets, MP3 players and the like – to capitalize on local Internet and computer-based services, while Japanese consumers are less likely to be able to exploit the full potential of the new technologies.

But it is axiomatic that the first and second phases of the digital revolution – the explosion of personal computer use and the development of business to business e-commerce – are but a precursor of what is to come. And it is at this level – the introduction of e-services, e-health and e-education – that anticipating the future becomes particularly problematic. Current balances of digital power and levels of computer use might well be a poor guide for what follows in the future. Culture (values, customs, and loyalties) is rarely factored into the IT equation. Instead, discussion focuses on technological questions – band-width, computer usage, connection rates, technological diffusion. But culture matters in business and in national economic development, perhaps more than most commentators recognize or admit. It may well be so with the digital revolution. To this point, the IT era has been dominated by technological change and an enormous amount of excitement and advertising. As enthusiasm for the dot-coms wanes and as investors, banks and corporations turn their attentions away from infrastructure and basic communications, the next stage of the IT revolution will focus on implementation and adoption of a wide range of digital services and communication technologies.

It is here, in the intersection of national culture and technological change, that some critical issues emerge. Put simply, some countries are better situated to respond to the opportunities of the digital revolution and to translate technological abilities and resources into social, economic and cultural opportunities. There are certain preconditions that will determine the two critical elements in the expansion of the information age: technological abilities and the willingness of the citizenry and its government to shift to a digital environment. It follows logically that some countries have the necessary preconditions for continued and future success and that others do not. And one of the key elements in any major economic or technological revolution is that the certainties of the past melt into ambiguities. As the digital revolution plays out, some countries will move ahead on comparative national scales and others will fall behind. The key question here is where Japan will be in the coming era of international competition.

Conventional wisdom has it that the overwhelming cultural advantage rests with the United States of America, although closer examination reveals that this may not be so. A great deal has been made about America's formidable lead in creative output – a critical element in the digital age – but one that is not well-understood and may be overestimated. The United States has done an excellent job of unleashing the creative and technological energies of a small portion of its citizens – and of many immigrants drawn to the USA by educational or investment opportunities. Time may well reveal that the United States's greatest advantage in the digital age was not creativity, but rather the two-fold combination of a risk-oriented investment system (legitimized high risk speculation on a staggering scale) and the hyper-competitive corporate world that exists at the cutting edge of financial uncertainty and product innovation. Silicon Valley and its regional counterparts are zones of financial risk-taking; venture capital firms have contributed as much or more to the American digital revolution than the country's universities and technical schools. Many of the companies have failed, and many more will tumble ignomiously in the coming years, but the handful of success stories propel the industry along. But severe difficulties in the stock-market, presaged in the summer-fall 2000 assault on Internet and technology stocks, would quickly dry up the pool of venture capital and send it scurrying to other fields (particularly bio-technology, which may soon emerge as an even more dominant economic area than information technology).

Will the American advantage hold? That is by no means assured. Many of the brightest minds in America's digital revolution are not Americans, but rather recent migrants from overseas. Some of these – a trickle at present – are returning home, pulled by culture and loyalty, or by juicy incentives offered by national governments. If that trickle turns to a flood, America will have provided the risk capital and financial opportunity for the development of the global Internet and digital economy. The USA's mediocre education system remains seriously deficient in scientific and technical areas, and the same liberal culture of personal choice and individuality which allegedly underlies the Internet explosion also contributes to a long-term crisis. American students routinely opt out of difficult science-based courses early in their high school years, creating a generation of digitally engaged and technologically illiterate people who lack the deep and sophisticated understanding necessary to sustain digital industries. The United States responds by expanding the number of visas available to overseas technicians and scientists, but this is not a long-term solution. If technological sophistication remains a key underpinning of the digital revolution, as it most certainly will, look for the United States to lose its lead and begin to slip behind other countries.

An examination of the foundations of continued and future success in the digital revolution suggests that, although they face disadvantages in some critical technological areas, Asian nations enjoy considerable opportunities. Without straying down the contentious path of arguing for Asian 'uniqueness', a hotly debated relic of the days of the 'Asian miracle', the reality is that there are fundamental and identifiable differences between leading Asian industrial nations and their counterparts in Europe, North America and Australasia. American individualism does not have a counterpart in Japan or China, although Finland, Canada and New Zealand share much in common with the United States. Asian education systems function very differently from those in North America, with one favouring national guidance and the other encouraging student and parental choice. National attitudes toward the role of government and the relationship between business and government vary dramatically. Each of these areas, and several others, may well prove critical to the medium-term unfolding of the digital revolution.

Some countries will exploit the Internet and the information age better than others. America has done this with e-commerce and experiments with entertainment on the Internet, and Japan has done so with m-commerce and with the commercialization of digital content. Nations take radically different approaches to the availability of information. Canada and Finland encourage open international access; China and Myanmar fear the ideological and political ferment will follow digital data into their country and seek to establish information firewalls to keep unwanted information at bay. The most powerful uses of information technology – for education, health care delivery, industrial processes, and government services – remain substantially untested. The world is, in fact, only at the outer edge of the full digital age and a great deal remains unknown about how the people and nations of the world will respond to the opportunities and challenges that lie ahead.[10]

Continued success at the cutting edge of technological innovation requires a first-rate education system and a commitment to the inclusion of technology in the classroom. Most western nations (Scandinavia and northern Europe are exceptions here) make proud noises about wired classrooms but then produce education systems of uncertain value, with teachers saddled with a baffling range of educational responsibilities and social pressures. North America's education system has failed to provide a steady stream of technologically proficient and creative graduates able to meet the needs of the digital age. Both Canada and the United States, as a consequence, import large numbers of scientifically trained professionals, providing a short-term response to a pressing national need. This flow of overseas personnel is unlikely to continue uninterrupted. One of the realities of the digital age is that trained professionals can work anywhere they wish – including in their home countries. Over time, the North American education system may prove to be a major Achilles heel in the Canadian and American attempt to maintain digital momentum.

Asia offers a different model. Japan's highly centralized, government-controlled education structure lends itself well to the introduction of technological teaching and the nation's demanding and high pressure, science-based high school program produces graduates with a substantial level of technological literacy. Other Asian nations – Singapore, Taiwan, Hong Kong and China – have highly centralized education systems that place a very strong emphasis on scientific education. While much remains to be done – North American classrooms and libraries are more 'wired' than their Asian counterparts – the reality remains that the underpinnings of a successful digital economy require training scientific personnel. The work ethic of the Asian student, the ability and willingness of governments to dictate educational content and standards, and broad societal support for a rigourous, demanding school system will likely ensure that these countries will continue to outperform North American and most European countries in the education of skilled scientific professionals and that the gap will widen over time.

Consider one possibility, as applied in two countries. Developing a basic appreciation for digital technology is a basic requirement for scientific and commercial competence in the 21st century. Educators would likely agree that the introduction of digital literacy programmes in elementary and high school would be beneficial for all students and would provide an underpinning for more advanced scientific study. In Japan (or other Asian countries), the national government has the authority – and sufficient public support – to implement such a strategy. It might take a year or two to design the curriculum and implement the changes, but a fairly rapid reorientation of the education system to accommodate digital needs is possible. Imagine the same approach in the United States, a country with strongly divided responsibilities for education and with strong local school boards. In the very best school divisions – with lots of money, parental support and high standards – such changes could proceed quickly (Subject to the critical realization that North American parents are very conscious of the academic success and progress of their children and tend not to tolerate failure or low grades. This has contributed to a lack of discipline in the classroom, limited teacher influence, and grade inflation.). In poorer districts – the majority of the country – facilities are already woefully inadequate, school performance is abysmal and the prospects are slight of parents and local governments funding and supporting major technological additions to the curriculum.

Education will be critical in determining the future of national capabilities in the information age, and Asian nations have a decided advantage. Japan, with one of the world's most tightly controlled and demanding high school systems, is exceptionally well-placed to ensure that its graduates are able to participate fully in the information age. It is not yet certain if the national government will push digital concepts this hard or in this particular direction. But the potential exists for Japan to use its education system to sustain and add to its national advantages in the digital age.

Much has been made of declining public support for government-led economic development in Japan, hardly a surprising development given the bursting of the bubble and the lengthy recession of the 1990s. External observers have overstated the impact of the economic difficulties on the average Japanese person. Per capita incomes have remained high – and the value of the yen on international currency markets has made Japanese very wealthy in global terms. Moreover, the country's economy has remained innovative and creative. Behind the seemingly endless stories of the failures of major department stores, insurance companies and banks lies a significant attempt to restructure the Japanese economy. Although the effort is not precisely what foreign business people want (why it is that western observers still feel so confident and arrogant in telling Asian governments what they should do and how they should manage their economies?), the reality is that major, government-led change is underway. Banking and insurance have been de-regulated, significant changes have been introduced for investment rules, and the government of Japan has begun the difficult process of laying down an administrative framework for the contemporary information technology sector.

Earlier enthusiasm for the capitalist development state – the powerful Asian union of government and major corporations, embodied in the phrase Japan Inc. – and for substantial government guidance of the economy has waned, inside and outside Japan. The government's handling of the bubble economy discredited state intervention in many eyes, and the power of the fabled Ministry of International Trade and Industry appears to have waned. Stories of political corruption and the cancerous torment of the Liberal Democratic Party through the 1990s gave the impression of a country in disarray and without guidance. Current displeasure with the Japanese government overstates the administrative and economic difficulties of the LDP and of Japan. The core of the nation's economic success did not rest in one political leader or in one government ministry. Instead, national competitiveness and innovation emerged out of the Japan-first ideology of the post-war period and the realization by Japanese business and consumers that collaborative action is in the collective best interests.

The broader model, rather than the individual actors, has not been discredited. At the national and prefectural level, governments continued to lay down the infrastructure for continued economic prosperity. While some investments are clearly political boondoggles – the vast wastage of government money on construction projects reeks more of political patronage that wise economic stimulation – many are well-suited to the country's effort at international competitiveness. Huge industrial parks, trade centres, and technological research units are clearly designed for the economy of tomorrow more than that of yesteryear. Through such organizations as the Keidanren (Federation of Economic Organizations, which merged in the fall of 2000 with the Nikkeiren (Federation of Employers Association), Japanese business engages in a sustained, intense dialogue with government about the country's prospects. Business exerts considerable authority, not as cacophony of competing voices (as is typically the case in the western nations) but as a strategically vital interest group with obvious commitments to national prosperity.

Japan, along with the other leading Asian economies, has passed through turbulent waters in recent times, and it is uncertain as to how well the country will perform in the future. The 1990s boom in the United States, much like the even more dramatic Bubble Economy in Japan in the 1980s, created an aura of invincibility and resulted in unwise and unsustainable straight-line projections about America's economic future. It is unclear which system – the *laissez-faire* American model or the government–business cooperative model – holds the best opportunities for long-term success. What is evident is that Japan, even in the midst of significant deregulation, holds to the view that business and government should work closely together in charting the country's future. The combination of the digital revolution and an era of economic globalization has challenged the efficacy of the Japanese approach, but it has yet to be proven a failure. Recession notwithstanding, Japan continues to reorient its economy toward the information age and appears to be at the point where political rhetoric, business needs and government actions converge. When this has happened in the past, Japan has raced to the fore. It may well do so again.

Analysts of the digital revolution have argued that the new economy will be the antithesis of the centrally managed and government-influenced Asian model. It will, advocates, claim, be free-wheeling, innovation-based, highly variable and very uncertain. Commentators talk of business working in Internet time and of the rapid changes in corporate and economic fundamentals in the face of digitally-mediated international competition. Enthusiasm for the *laissez-faire*, competitive market should be contained by basic realities. Infrastructure and massive investments underlie this revolution. Trillions of yen have been invested in the factories that provide the semi-conductors and exceptionally specialized components that drive the digital age. Telecommunication firms gamble millions of dollars on band-width and local connection services. Governments face enormous costs in attempting to bring the benefits of the information economy into the hospitals, schools and administrative offices of the nation. Small companies, fast-moving, creative and irrepressibly optimistic, may define the parameters of what is possible, but carefully planned, extremely costly and commercially and politically viable investments are necessary to convert technological dreams into practical implementations.

To date, with the digital age still in its formative and speculative phases, the baton has rested in the hands of the speculator, the inventor and the promoter. As the dot coms discovered through 2000, however, there is a limit on corporate and investor patience. There is a time when energy, enthusiasm and zeal is sufficient to sustain optimism and cash flow. But at a critical juncture, the challenge of turning visions into reality fall to the usual combination of regulators, politicians and business leaders, the ones who determine and shape the nation's economic future. The information technology revolution appears to be hitting this stage, the transition from infancy to maturity and from bold visions to practical implementations. Governments stand to be more important, not less so, in determining the future path of technological development and in ensuring that the many tested and practical applications of digital technologies – many of which fall well within the public realm – are actually brought into service.

The Japanese model of government-business cooperation might serve the country well in subsequent phases of the digital revolution. While this is not assured, it is clear that future progress rests on the ability and willingness of governments and business to identify common cause and to identify shared objectives. Japan has been very good at this in the past and the lines of communication remained open and functioning through the Bubble Economy, the recession and into the 'Big Bang' of regulatory reform. It may well be that in Japan, and elsewhere in Asia, the model of government-business engagement could well propel the country and region ahead of more *laissez-faire* states, where government enjoys minimal standing and exercises less direct influence over the pace and nature of economic change.

Japan may also enjoy a pivotal advantage at the individual level, most notably in the ability of the Japanese people to place national objectives ahead of personal needs. The evidence of the last decade clearly shows that consumer resistance is the greatest single barrier to national success in the information revolution. The Japanese have a well-known reputation as early adopters in terms of technology; they leap at new commercial products, try out new services, and respond quickly to interesting developments. The phenomenal response to DoCoMo's i-mode phone is but the most remarkable example of this deeply ingrained trend. While it is too early to see how widely this carries, into e-commerce for example, the Japanese have an enviable record for adapting to new technological opportunities and trying out new services.

How countries respond to the broad and sweeping potential of the information revolution will be determined, in large measure, by their continued adaptation to new technology-based delivery systems. By the end of the 20th century, the world had seen only the easiest and most obvious roll-outs, in direct sales, e-based advertising, and information sharing. Most of these ventures remain money-losing propositions, although companies like Amazon.com and e-Bay have attracted sizable client lists. How people respond to the next and more critical series of developments is crucial to the information revolution realizing even a significant part of its potential. Will people accept medical services over the Internet, including imbedded heart devices controlled remotely by medical practitioners? Will they be comfortable when 50 per cent of their children's school time is devoted to Internet-based classes, and will they abandon, in large numbers, traditional universities and colleges in favour of home-study via computers? It is not yet known if the general public will be comfortable dealing with local, regional and national governments electronically and if they will overcome their inherent suspicion about privacy of information on the Internet.

As these issues unfold – and these are the critical questions that will determine if the information revolution becomes central to contemporary life or stays, for most people, on the margins – it is likely that national cultural imperatives will come into play. In many western industrial nations, the primacy attached to individuality and privacy serves as a brake on the acceptance of new technologies. Similarly, the prerogatives of the western service culture have, to date, slowed the acceptance of e-health and many e-services. While technologically sophisticated groups – investors, for example – have flooded to Internet-based stockmarket web-sites, the average citizen appears less likely to move in this direction. Countries, like Japan, that are used to more centralized control and that are used to being told by their governments how and where services will be provided, have the potential to respond far more favourably to technologically based delivery. Government decisions to provide a portion of schooling over the Internet, to require the integration of the Internet into college and university instruction, or to limit the provision of medical services in remote regions to e-health initiatives are far more likely to be accepted, if not embraced, in Japan than in many other industrial nations.

The convergence here of the potential willingness of the Japanese people to accept government services as provided – rather than to demand those services in a set format or delivery system, as would be the case in North America – and the need for government leadership in the implementation of next-generation digital initiatives is critical. E-health is possible, and current and future innovations have the potential to change radically the prevention, assessment and treatment of disease. E-education and E-government models are well in hand, and hardware and software exists to make the implementation of services in these areas relatively quick. What is unknown is the degree of public acceptance and the willingness of citizens to accept government provided services over the Internet or in some digital format. Evidence to date from North America – where acceptance of E-health lags far behind the testing of viable digital delivery systems – illustrates the fundamental truism of the information revolution: what people will accept is ultimately more significant than what one technology can provide.

Because the range of acceptance in Japan is significantly broader than in North America and other western industrial regions, the country has a formidable advantage, one that has not yet been fully developed. Japanese citizens are not as demanding of their government, and expect far less personal input into the provision of government services. Dealing with government offices is complex, slow and resistant to citizen protest. Medical services are doctor, not patient, centred and place remarkable demands on the ill and their families. Education is rigid and formalized and does not accommodate local input. While the formality and rigidity has long been seen by outsiders as a major liability, it might well transpire that the Japanese attitude to government is a critical competitive advantage.

Government agencies have the potential, more so in Japan than in most western countries and in common with many Asian nations, to shift services and delivery systems to digital formats. With a very high rate of Internet use (particularly since the advent of the i-mode phone), most Japanese citizens have routine and inexpensive access to digital communications systems. Add to this a general acceptance of the dictates of government – something that simply does not exist in North America, for example – and Japan has a formula for the quick and effective roll-out of government programmes using information technologies. The country and its Asian neighbours have unique opportunities to take global developments in key service areas and to apply them to the schools, hospitals and government offices, confident that the citizenry at large will be generally accepting of the transition. Other countries salivate at the prospect of a compliant, technologically flexible target population. Japan seems temperamentally, technologically, economically and socially predisposed toward the imperatives of the network society – a population tied together by the ready and accessible exchange of information through technologically mediated systems.

Scholarly work on the intersection of Internet technology and societal values and changes is at a very early stage. In his sweeping study, *End of Millennium*, Manuel Castells addressed the question of Japan's reaction to the potential *Johoka Shakai* (information society), a concept which he asserts originated in Japan. Castells argues that the government –industrial emphasis on collaborative economic development sparked a rapid technological response to the prospects of the digital revolution, particularly on the manufacturing side, and unwittingly launched a major restructuring of Japanese society. He states, 'Japan built a new mythology around a futurological view of the information society, which actually tried to replace social thinking and political projects with images of a computerized/telecommunicated society, to which were added some humanistic, pseudo-philosophical platitudes'. The resulting 'culturally/historically specific model of the information society' inevitability 'came into contradiction not only with the technocratic blueprints of an abstract social model, but with the institutional and political interests of its procreators. Furthermore, after Japan bet its entire technological and economic development on the informational paradigm, the logic of the state came into contradiction with the full blossoming of this paradigm'.[11]

Castells' work on the network society speaks to the evolution of a population toward an information-rich system of connections, interactions and decision-making. He foreshadows a world dominated by the exchange of information, in which economic relations have been substantially reconfigured as a consequence of the advent and availability of information technology. In his review of Japan as a network society – he argues Japan was such a society in the 1990s – Castells suggests that the new order stands in contradiction with the structures, institutions and values of post-World War II Japan. He advances eight main reasons why contemporary Japan as a network and information society stands apart from its historic condition:[12]

1. Corporate globalization has weakened the power of the Japanese government and exposed the short-comings of the Japanese development state;
2. Liberalization of trade and the deregulation of society (particularly the telecommunication sector) has significantly reduced the levels of power available to the national government, giving the population greater freedom in the process.
3. The shortcomings of Japanese research, particularly the bureaucratic university system, have limited the country's ability to capitalize on its technological advances and to remain at the forefront of scientific innovation. In particular, the government's commitment to a superconductor based information society was not offset by entrepreneurial innovation in the critical area of computer-mediated interactions.
4. The rigidities of the life long employment system interfered with the required flexibility of a true information society and slowed innovation across the country.
5. Japanese traditional identity is being supplanted rapidly by 'the culture of real virtuality', as seen in computer games, *anime*, and various other technological and cultural responses to traditional Japanese society. The multiplicity of images and identities will, he argues, over-ride the connective power of traditional Japanese values and customs.
6. The long-standing Japanese commitment to nation and nationality are under pressure from a variety of social movements, many of which have been empowered by the communication pathways of the information revolution. Japanese social homogeneity is being exposed as a coordinating myth, not a reflection of Japanese values and commitments.
7. The network and information society in Japan is actively engaged in the redefinition of 'civil society' and, in the process, is

    intensely critical of the cronyism and corrupt national political system. The struggle to redefine politics in the interests of society, instead of the interests of the politicians, will reconfigure Japanese institutions.

8. Japan's response to the Internet (the first edition of *The End of Millennium* was produced in 1998, before the *keitai* revolution took hold) demonstrates that the information society is mounting a concerted attack on the status quo. The internet, email, ecommerce and other innovations have, he argues liberated hitherto isolated individuals and groups to assert a strong role in the new Japan.

Many of the conclusions in this book mesh with the analysis advanced by Castells, although we argue that the strengths of post-World War II Japan remain clearly in evidence. Further, developments over the last five years have suggested that Japan is capable of adapting and responding quite quickly to new opportunities and threats. The government is, for example, embracing aspects of the information revolution, and has assumed a leadership role in developing practical applications for Internet-based technologies. The university system is in the midst of a major restructuring, as is the system of government-academic-private sector research. Japan, it seems, has learned from the problems of the past two decades. Life long employment may have been eroded, but there appears to be evidence that the cracks in the system have increased the attachment of individuals and corporations to the concept. There is yet to be compelling evidence that information technology is sparking a revolution in political activism and mobilization; one of the most dramatic applications of the technology to politics has come through Prime Minister's Koizumi's establishment of his weekly email newsletter to over 2 million subscribers. Major companies are responding to the potential of the Internet, rather than being overturned by upstart competitors. It is Softbank, not Sony and Toyota, that has seen its standing evaporate after 2001. Finally, the Internet has, as we have argued, proven to be a boon to Japan-ness, not a threat. As the largest nation-specific Internet system in the world, the Japanese network has erected linguistic firewalls to the world, and the country has not yet experienced the crumbling of cultural distinctiveness that many commentators forecast.

    What, ultimately, lies ahead for the digital revolution in Japan? The country is a world leader in key information technologies, including mobile telephony and Internet services, industrial and commercial use of robots, practical m-commerce models, and manufacturing capabilities in critical digital fields. Moreover, Japan has a compliant and accepting population that is responsive to government direction, a dynamic and innovative consumer culture, a long-standing fascination with new technologies and new products, a strong education system, and a deep thread of corporate innovation and international competitiveness. On the negative side, Japan started slow in the Internet race; government, in fact, refused for a time to acknowledge that the race was underway and only reluctantly and belatedly came to the conclusion that it was a contest worth entering. Infrastructure and the regulatory environment continues to stifle Internet-based initiatives and saddles the country with a serious burden in its attempt to remain active and competitive. Language is both a barrier and an advantage, ensuring that the country loses relatively little to international e-commerce (although localization may change this) and provides a wealthy and curious capital market within Japan for Japanese-based offerings.

Prediction in the field of information technology is fraught with difficulties. Nicholas Necroponte, one of the leading thinkers about the digital revolution, seriously over-estimated the impact of the Internet on the entertainment and news industries. Bookshelves groaned with the weight of hundreds of books and magazines forecasting the endless expansion of the Internet economy and the future prosperity of the dot.coms, assumptions and assurances which collapsed almost as rapidly as they emerged. The expenditure of millions of dollars by thousands of companies have brought precious few profits and a lot of failures. Convergence of technologies, particularly the Internet and television, and the implementation of micropayment schemes might alter the equation but returns in this area have also been meagre. Assessing national prospects is even trickier, for the collision of culture, politics, business, consumer interests and technology creates a tangled web of uncertainty.

Japan will be a formidable player in the digital revolution. It will probably expand, rather than relax, its dominant position in robotics and the development of the components of the information age. The i-mode/mobile Internet innovation is at its beginning and usage will grow dramatically in Japan and spread rapidly throughout the industrial and developing world. Japanese companies will be at the forefront of this transition, selling the phones and technological backbones that will connect the wireless world. As the slow-moving Japanese government is mobilized, it is reasonable to anticipate substantial reductions in regulatory interference and Internet costs, thus propelling a new wave of home computer-based Internet development. Look, in particular, for continued emphasis on the convergence of television, computers, mobile phones and other household appliances, as Japanese companies respond to their national consumers' deep curiousity and willingness to experiment.

If Japan chooses – and this option appears to be at least partially on the table – the country could well be the leading nation of the next major phase in the development of the Internet. Technological innovations will continue, but the major barriers have been overcome and issues of speed and bandwidth substantially addressed, at least in the laboratory. Japan has abundant pools of investment capital which could readily be tapped to finance a major expansion of Internet service delivery. Moreover, if the Japanese government capitalizes on the responsiveness of its citizenry and moves aggressively toward an e-service mode, it could move its administrative operations, schools and hospitals out of the 1960s and into the 21st century in quick order.

Stepping outside the rhetoric and hysteria surrounding the information revolution, the reality is that the world still has very little evidence of what the Internet and allied digital technologies are capable of doing. In North America, the most dramatic implementations of the Internet have been email, reservation systems, auction houses and, most significant of all, the Napster-led attack on musical copywrite and the very foundations of the music industry. But these are hardly revolutionary changes. Cassette tapes presaged Napster by 30 years, and online reservation systems have primarily served to eliminate jobs in the travel industry. Only a handful of people – most in pilot projects – receive medical care over the Internet. E-commerce has been far less successful and the corporate fall-out in this sector continues to expand. Digitally based education is growing, but largely by supplanting old forms of distance education (such as correspondence courses). The cumulative effect is significant, but hardly overwhelming.

Chances are quite strong that Japan will be among the first nations to push the digital model to its next logical step. The country's ready adoption of robotic machinery and the i-mode are good indications of national receptiveness. As to the commercial possibilities of the digital revolution, far greater potential is evident on Japanese trains, in classrooms, on street corners (and even, rather surprising, among bike riders) as young Japanese people fiddle with their i-mode phones and download services, information, games, music and other attractions. I-mode will likely do what promoters thought e-commerce would accomplish. Add to this the possibilities – more readily available in

Japan than most countries – for the potential widespread, government-directed use of digital services, and an image emerges of a country prepared to move beyond rhetoric and beyond the speculators' visioning into a truly digital age, where computers, chips, the Internet, and other digital devices both define and support the nation. Japan is not yet a networked nation. Japan.com remains a possibility rather than a reality. But underestimating the Japanese yet again might well prove to be a serious mistake.

Notes

Introduction: The Tip of the Iceberg

1. Terutomo Ozawa, 'The new economic nationalism and the "Japanese disease": The conundrum of managed economic growth', *Journal of Economic Issues*, vol. 30, June 1996, pp. 483–92.
2. 'Going for IT', *Nikkei Weekly*, 23 October 2000.
3. To follow the release of new products, see Akihabara Watch, a web-site maintained by the Asian Technology Information Program. See http://www.atip.or.jp/Akihabara/
4. David Freeman, 'Japan: the maybe restoration', *Forbes.com*, 21 Feb. 2000.
5. For an early overview see Philippe Perez, 'The Internet Phenomenon in Japan', *Communications et Strategies*, 1998, no. 32, p. 157.
6. There is a vast literature in this field. See, for example, Edward Malecki, *Technology and Economic Development: The Dynamics of Local, Regional and National Change* (New York: Longman, 1991). For a recent series of sectoral and national studies, see Been Steil, David Victor and Richard Nelson (eds), *Technological Innovation & Economic Performance* (Princeton: University of Princeton Press, 2002). See also Zaheer Baber, 'The Internet and social change: key themes and issues', *Asian Journal of Social Science*, 2002, vol. 30, no. 2, 195–8.
7. Tom Forester, *Silicon Samurai: How Japan Conquered the World's IT Industry* (Cambridge: Blackwell, 1993), x.
8. Among the most influential of these books were Chalmers Johnson, *MITI and the Japanese Miracle: The Growth of Industrial Policy* (Stanford: Stanford University Press, 1982); Ezra Vogel, *Japan as Number One: Revisited* (Singapore: Institute of Southeast Asian Studies, 1986).
9. Among the most compelling of this genre is Akio Morita, with Edwin Reingold and Mitsuko Shimomura, *Made in Japan: Akio Morita and SONY* (New York: Dutton, 1986). For a more comprehensive view of Japanese entrepreneurship related to digital technology, see Bob Johnstone, *We Were Burning: Japanese Entrepreneurs and the Forging of the Electronic Age* (New York: Basic Books, 1999).
10. A. Chandler, *Inventing the Electronic Century: The Epic Story of the Consumer Electronics and Computer Industries* (New York: New Press, 2001).
11. Eamonn Fingleton, Blindside: *Why Japan Is on Track to Overtake the U.S. by the Year 2000* (Boston: Houghton Mifflin, 1995).
12. E. Fingleton, *In Praise of Hard Industries* (Boston: Houghton Mifflin, 1999).
13. Adam Posner, 'Japan', in Been Steil, David Victor and Richard Nelson (eds), *Technological Innovation and Economic Performance* (Princeton: Princeton University Press, 2002), pp. 74–111.
14. Michael Porter, Hirotaka Takeuchi and Mariko Sakakibara, *Can Japan Compete?* (London: Macmillan – now Palgrave Macmillan, 2000), pp. 2–3.

15. See, for example, Mauel Castells, *End of Millennium* (New York: Blackwell, 2000), particularly ch. 4, 'Development crisis in the Asia Pacific: globalization and the State', pp. 220–55.
16. On the critical and fascinating semi-conductor industry, see the early study by John Tilton, *International Diffusion of Technology: The Case of Semiconductors* (Washington: Brookings Institution, 1971). See also Gene Gregory, *Japanese Electronics Technology: Enterprise and Innovation* (New York: John Wiley, 1986).
17. Koji Kobayashi, *The Rise of NEC: How the World's Greatest C&C Company Is Managed* (New York: Blackwell, 1991).
18. Marie Anchordoguy, *Computers Inc.: Japan's Challenge to IBM* (Cambridge: Harvard University Press, 1989)
19. Martin Fransman, *The Market and Beyond: Cooperation and Competition in Information Technology Development in the Japanese System* (Cambridge: Cambridge University Press, 1990).
20. Ken-ichi Imai, 'The Japanese pattern of innovation and its evolution', Nathan Rosenberg, R. Landau, and David Mowery (eds), *Technology and the Wealth of Nations* (Stanford: Stanford University Press, 1992), pp. 225–46.
21. Michael Cusamano, *The Japanese Automobile Industry: Technology and Management at Nissan and Toyota* (Cambridge: Harvard University Press, 1985); Michael Cusumano, *Thinking Beyond Lean: How Multi-Project Management Is Transforming Produce Development at Toyota and Other Companies* (New York: Free Press, 1998).
22. Michael Cusumano, *Japan's Software Factories: A Challenge to U.S. Management* (New York: Oxford, 1991), p. vii.
23. Manuel Castells, *The Rise of the Network Society* (New York: Blackwell, 1996), pp. 31–2.
24. See, among a vast array of books, Bill Gates, *Business @ The Speed of Thought* (New York: Warner, 2000), Michael Dertousos, *What Will Be: How the New World of Information Will Change Our Lives* (New York: HarperBusiness, 1998); and Nicholas Necroponte, *Being Digital* (New York: Vintage, 1996).
25. For an interesting sector by sector assessment of the Internet as it relates to the United States, see Brookings Task Force on the Internet, *The Economic Payoff from the Internet Revolution* (Washington: Brookings, 2001).
26. As this relates to post-secondary education, see David Noble, *Digital Diploma Mills: The Automation of Higher Education* (New York: Monthly Review Press, 2002).

1   Uneasy Steps: Japan and the Development of the Digital Society

1. Robert Grey, 'Turning Japanese', *Marketing*, 21 Sept. 2000, pp. 34–35.
2. For an excellent overview of the consumer electronics and personal computer sectors, with considerable emphasis on Japan (particularly Sony), see Alfred Chandler, *Inventing the Electronic Century: The Epic Story of the Consumer Electronics and Computer Industries* (New York: Free Press, 2001).
3. Anchordoguy, Marie, *Computers Inc.: Japan's Challenge to IBM* (Cambridge: Harvard University Press, 1989).

4. Ibid.

5. Ibid.

6. Anchordoguy, Marie, *Computers Inc.: Japan's Challenge to IBM* (Cambridge: Harvard University Press, 1989), p.122.

7. Anchordoguy, Marie, *Computers Inc.: Japan's Challenge to IBM* (Cambridge: Harvard University Press, 1989), p.143.

8. Japan Information Processing Development Centre, *Informatization White Paper*, 1995.

9. Charles Bickers, 'Banking on the robot evolution', *Far Eastern Economic Review*, vol. 163, no. 47, 23 Nov. 2000, pp. 38–42.

10. Irene Kunii, 'Robots', *Business Week*, issue 3724, 19 Mar. 2001.

11. Sam Joseph, 'Robots R Us' – Move over Aibo', J@pan Inc., Feb. 2002, p.15.

12. 'Sony rolls out futuristic Aibo capable of wireless PC link', *The Nikkei Weekly*, 12 Nov. 2001, 17.

13. Sam Joseph, 'Robots R Us – move over Aibo', J@pan Inc., Feb. 2002, p.17.

14. 'Matsushita's robvot loves air hockey', *The Nikkei Weekly*, 20 Oct. 2001.

15. 'Consumers ready for home robots', *The Nikkei Weekly*, 13 Jan. 2003, p.1.

16. 'Humanoid robots taking big strides in evolution', *The Nikkei Weekly*, 23 Dec. 2002, p.11.

17. Sam Joseph, 'Robots R Us – move over Aibo', *J@pan Inc.*, Feb. 2002, p.19.

18. Charles Bickers, 'Banking on the robot evolution', *Far Eastern Economic Review*, vol. 163, issue 47, 23 Nov. 2000, pp. 38–42.

19. Such as the iBox, to permit web-surfing via television, a technology which has yet to find a sizable audience. See Yvonne Lee, 'Japan Computer Corp. unveils Internet device', *InfoWorld*, 26 Feb. 1996.

20. Yves Lehmann, 'Japan on the move', *Telecommunications*, vol. 29, no 3, Mar. 1995.

21. Michael Martin, 'When info worlds collide', *Fortune*, vol. 134, no. 8, 28 Oct. 1996, pp. 130–6.

22. Charles Dogson, 'Who needs imode', *Communications International*, Jan. 2001.

23. James Niccokai, 'Japan's i-mode urges chipmakers to think mobile', *InfoWorld*, Vo. 23, no. 9, 26 Feb. 2001.

24. Terho Uimonen, 'U.S., Japan set for broadband competition', *InfoWorld*, vol. 23, no. 6, 5 Feb. 2001.

25. Chang-tseh Hsieh, 'Japan's quest for global leadership in IT: the impact on US computer companies', *Industrial Management*, vol. 94, no. 2, 1994, pp. 23–28.

26. *Japan Almanac 2000*, pp. 30–32.

27. 'The information technology gap', *Far Eastern Economic Review*, 30 June 1994. See also, 'Leading edge', *Asiaweek*, 12 May 2000.

28. For one of the many commentaries on this issue, see David Hayes and Martyn Warwick, 'Building bridges to the future', *Communications International*, vol. 24, no. 4, Apr. 1997, pp. 8–16.

29. Irene Kunii, 'Broadband in Japan: the brawl begins', *Business Week*, 12 Mar. 2001.

30. 'Japan adjusts to the broadband era', Asiamoney, Dec.200/Jan. 2001.

31. 'Content providers rush to reap riches from broadband', *Nikkei Weekly*, 4 Nov. 2002.

32. Japan Information Service Industry Association, *The IT Services Industry in Japan 200* (JISIA, 2000), p. 63.

33. Ibid., 66.
34. 'IT companies form group to offer graduate courses', *Nikkei Weekly*, 13 Jan. 2003.
35. 'Cheap wireless LANS emerging to challenge 3G cell phones', *Nikkei Weekly*, 29 April 2002; 'Japan in Focus', *Nikkei Weekly*, 8 July 2002, p. 7.
36. S. Fukukawa, 'Awakening to the IT revolution', *Japan Quarterly*, 1 October 2002, vol. 47, no. 4. On how this relates to technological change, see Marie Anchordoguy, 'Japan at a technological crossroads: does change support convergence theory,' *Journal of Japanese Studies*, 1997, vol. 23, no. 2, p. 363.
37. For a review of the forces buffeting Japan at this time, see J. Eades, *et al.*, *Globalization and Social Change in Contemporary Japan* (Melbourne: Trans Pacific, 2000), pp. 1–16.
38. I. Ieo, 'Being digital in Japan: The Current Internet Fever and Multimedia in Japan', *Journal of Japanese Trade and Industry*, 1 Nov. 1995, vol. 14, no. 6; I. Ieo, 'The state of digitization in modern Japan: the road to an advanced information network society', *Journal of Japanese Trade and Industry*, 1 Jan. 1997, vo. 16, no. 1 offers early views on this issue.

2   Japan.com: Government and the Promise of the Internet Society

1. Eric Lim, 'Japan as number two', *Far Eastern Economic Review*,' 3 Feb. 1994.
2. Kaoru Morishita, 'A closer look – IT start-ups: Hokkaido hot spot', *Nikkei Weekly*, 14 May 2001.
3. Japanese Ministry of Education, 'Sangakurenkei now: partnership between universities and industry', Apr. 2002, p. 4.
4. 'Academia builds high-tech ties with private companies', *The Nikkei Weekly*, 22 Oct 2001, 3.
5. Ibid.
6. Bruce Lambert, 'Building innovative communities: lessons from Japan's science city projects,' European Institute of Japanese Studies, Working Paper 107, Nov. 2000.
7. General View of Kansai Science City (http://www.mlit.go.jp/crd/daisei/daikan/gaiyo-e.html).
8. 'Researchers seek to put "Internet car" on road', *Japan Times*, 6 Mar. 2001.
9. This section is drawn largely from Ken Coates, 'Back in the race: Japan and the internet', *Japan after the Economic Miracle: In Search of New Direction*, P. Bowles and L.Woods (eds), Kluwer Academic Press, 2000, pp. 71–84.
10. 'Wiring Japan', 5 May 1994, Wired (http:muhu.cs.helsinki.fi/mailing_lists/pointers/msg00035.html).
11. Ibid.
12. Ibid.
13. See Jun Murai profile, http://korea.park.org/Japan/Theme/sense/profiles/murai.html
14. Mariko Tomiyama and Yuko Maeda, 'Internet and Japan', 1994, (http://muhucs.helsinki.fi/mailing_lists/pointers/msg00036.html
15. Jun Murai, http://www.isco.org/isco/general/trustees/murai/shtml
16. For an excellent historical overview of the development of the Internet and, in particular, the public/private interface in North America, see Janet

Abbate, 'Government, business and the making of the Internet', *Business History Review*, 2001, vol. 75, no. 1, pp. 147–76.

17. 'Japan's Internet Tangle', *The Economist*, 15 July 1995.
18. 'An Interview with Jun Murai', *Science and Technology in Japan*, 199, vol. 18, no. 70, p. 29.
19. *For a review of the ISP situation*, see Daniel Grunenbaum, 'Internet service providers in Japan', *The Journal*, Jan. 1998, vol. 135, no.1 .
20. Bob Johnstone, 'Wiring Japan', *Wired*, Feb. 1994.
21. For an overview of this rapid expansion, see 'Internet fever arrives in Japan', *Research Technology Management*, 1996, vol. 39, no. 4, p. 2.
22. 'Internet global growth rate', *International Data Corporation*, 23 June 1997.
23. 'Orienteering: how Canadian entrepreneur Rogwer Boisvert found Internet fame and fortune in the scattershot world of high-tech Japan', *Canadian Business*, 2001, vol. 74, no. 12, p. 42.
24. Roger Boivert, 'Doing high technology business in Japan', presented sponsored by Ladner Downs and Deloitte Touche, Vancouver, 27 May 1998.
25. 'Deregulation, foreign incursions lead to new connections', *Japan Economic Almanac* 2000, p. 29.
26. In January 2001, the Ministry of International Trade and Industry's name was changed to the Ministry of Economy, Trade and Industry (METI).
27. Forest Linton, 'The digital forest', *Computing Japan*, Jan. 1997. (http:cjmag.co.jp/magazine/issues/1997/jan97/forest/html).
28. 'Japan's economy: Asia's so slow express', *The Economist*, vol. 357, no. 8195, 4 Nov. 2000, pp. 75–77.
29. 'Mori starts extra diet with e-Japan pitch, *Japan Times*, 22 Sept. 2000.
30. 'Full text of Prime Minister Yoshiro Mori's policy speech given to the 150th Diet session Thursday', Japan Times, 22 Sept. 2000.
31. 'Whispering reform', *The Economist*, vol. 342, no. 7999, 11 Jan. 1997, pp. 19–21.
32. Jim Rohwer, 'Japan goes Web crazy'. Fortune, vol. 141, no. 3, 7 Feb. 2000, pp. 115–18.
33. David Moschella, 'A modest proposal for Japan', *Computerworld*, 7 Sept. 1998.
34. Vanessa Cornu and Stephen McClelland, 'Japan: still wrestling with change', Telecommunications, 10. Oct. 1995, vol. 29, no.
35. 'Stimulus policy should continue', *Japan Times*, 21 July 2000.
36. 'Mori starts extra diet with e-Japan pitch' *Japan Times*, 22 Sept. 2000; 'Mori taps tech sector for IT advice', *Japan Times*, 22 Dec. 2000.
37. 'State adopts national info-tech strategy', *Japan Times*, 28 Nov. 2000.
38. 'Ministries apply for billions for IT', *Japan Times*, 7 Oct. 2000; '11 trillion yen plan gets official nod', *Japan Times*, 20 Oct. 2000.
39. Japan Information Service Industry Association, The IT services Industry in Japan 2000 (JISIA, 2000), 43.
40. Report from the IT Strategy Council, 27 Nov. 27 2000.
41. 'Cabinet approves e-commerce bill', *Japan Times*, 21 Oct. 2000.
42. IT vouchers, round 2', *Japan Times*, 27 Sept. 2000.
43. Michael Mahoney, 'Japan passes e-commerce catch-up bill', *E-Commerce Times*, 13 Nov. 2000; http://www.ecommercetimes.com/news/articles2000/001113-1.shtml

44. 'Bill calls for advanced telecom network', *Japan Times*, 20 Sept. 2000.
45. For one view of the potential of the IT initiatives in Japan, see T. Howell, 'The Internet as cure for Japan's Ills', *World Trade*, 2001, vol. 14, no. 7, pp. 42–3.
46. Brad Glosserman, 'No field of dreams', *Japan Times*, 25 Oct. 2000.
47. 'The road to e-future: panel promises five year plan to put Japan in IT lead', *Japan Times*, 23 Jan. 2001.
48. 'Japanese e-government and e-commerce', 17 Dec. 2001. (Internet source?)
49. For a general history of the role of NTT in national economic development, see M. Anchordoguy', Nippon Telegraph and Telephone Company (NTT) and the Building of a telecommunications industry in Japan', *Business History Review*, 2001, vol. 75, no. 3, pp. 507–41.
50. 'Attempt to increase competition', *Japan Times*, 10 Dec. 2000
51. Fiona Haddock, 'Telecoms in Asia' *Asiamoney*, Dec. 2000/Jan. 2001, p. 21.
52. Ibid.
53. 'State want fast L-mode introduction', *Japan Times*, 3 Feb. 2001; 'Ministry postpones decision on NTT Internet service', *Japan Times*, 1 Feb. 2001.
54. 'Can e-Japan make leap from paper to reality', *Japan Times*, 16 May 2001.
55. 'Japan's e-signature Law', *ECOM Journal* No.1, p. 51.
56. Japan Information Service Industry Association, *The IT Services Industry in Japan*, Mar. 2000, p. 43.
57. 'Can E-Japan make leap from paper to reality', *Japan Times*, 16 May 2001, 1.
58. 'Four-nation survey confirms growing acceptance of online health information', *Medical Marketing and Media*, 2002, vol. 37, no. 7, pp. 10–12.
59. Japan has some 3,000 occupied islands, many with very basic and minimal health care services.
60. For an example of how this relates to the field of oncology, see H. Mizushima *et al.*, 'Japanese experience in telemedicine in oncology', *International Journal of Medical Informatics*, 2001, 61(2–3), pp. 207–15. See also K. Matsumura *et al.*, 'A network system of medical and welfare information service for the patients, their families, hospitals, local governments, and commercial companies in a medical service area', *Journal of Medical Systems*, 2002, 26(3), pp. 249–54.
61. Ken Belson, 'In search of Doctor-san', *Business Week*, vol. 3711, 11 Dec. 2000.
62. Takashi Takahashi, 'The present and future of telemedicine in Japan', *International Journal of Medical Informatics*, 2001, vol. 61, no. 2, p. 131. See also 'Medicolegal topics on the Internet in Japan', *Journal of Telemedicine and Telecare*, 2000, vol. 6, no. 1.
63. 'The emergence of e-health in Japan', *Scrip Magazine*, 2002, no. 110, p. 21.
64. 'Ministry plans e-voting legislation', *Japan Times*, 26 Jan. 2001.
65. 'Okayama municipality blazes trail with e-voting', *Nikkei Weekly*, 1 July 2002, p. 5.
66. C.S. Thompson, 'Enlisting on-line residents: expanding the boundaries of e-government in a Japanese rural township', *Government Information Quarterly*, 2002, 19(2), pp. 173–88.
67. 'Governments slowly begin closing gap between rhetoric and reality as they push to create eGovernment, according to second annual global accenture study': http:www.accenture.com/xd/xd.asp?it=enweb&xd=_dyn/dynamic pressrelease_244.xml

68. 'I-briefs', *Far Eastern Economic Review*, vol. 8 Feb. 2001, 164, no. 5.
69. 'Prime Minister's Policy Speech', *Japan Times*, 8 May 2001.
70. 'Full text of Koizumi's policy speech to the Diet', *Japan Times*, 28 Sept. 2001
71. George Nishiyama, 'Millions subscribe to Japanese PM's e-mail magazine', *Globe and Mail*, June 2001.
72. www.kaneti.go.jp/m-magazine.hyou.0712_danjyo.html.
73. L. Tkach, 'Politics'Japan: Party Competition on the Internet in Japan', *Party Politics*, 2003, vol. 9, no. 1 (2003), pp. 105–23.
74. Despite record high levels of unemployment, the shortage of people to work in Japan's engineering and programming sectors came to 12,400 in February 1999 according to a survey. Tetsushi Kajimoto, 'Foreign workforce moving on up: dearth of IT workers prompts Third World recruiting efforts', *Japan Times*, 1 Jan. 2001.
75. Takemochi Ishii, 'Open network at core of Japan's IT strategy', *Japan Times*, 1 Jan. 2001.
76. For one view of opportunities for Internet-based education, see Fumie Kumagai, 'Possibilities for using the Internet in Japanese education in the information age society', *International Journal of Japanese Sociology*, 2001, no. 10, pp. 29–44. For an early overview of the possible impact of the Internet on Japanese education, see T. Morris-Suzuki and Peter Rimmer, 'Cyberstructure, society and education: possibilities and problems in the Japanese context', paper presented to the National Institute of Multimedia Education, International Symposium, 1999.
77. *Asian Wall Street Journal Weekly*, 2000, vol. 22, no. 36, p. 2.

3 The Keitai Revolution: Mobile Commerce in Japan

1. The credibility and reliability of e-commerce payment systems was a crucial element in the slow development of Japanese e-commerce. For a useful commentary on this issue, see Katsuhiro Iseki, 'Approaches to ensure credibility of EC with consumers', *ECOM Journal*, No. 1 (www.ecom.or.jp) and Tsuneo Matsumoto, 'Development of electronic commerce by building consumers' confidence', *ECOM Today*, No. 7 (www.ecom.or.jp).
2. The risks of under-estimating Japan's economic adaptive capabilities are revealed in recent surveys which indicate that e-commerce, particularly in the B2B sector, is expanding very rapidly. See Paul Greenberg, 'Japanese e-commerce set to explode', *E-Commerce Times*, 21 Jan. 2000.
3. The government of Japan is determined to overcome regulatory and logistical barriers to e-commerce and m-commerce. See Michael Mahoney, 'Japan passes e-commerce catch-up bill', *E-Commerce Times*, 13 Nov. 2000. (www.Ecommercetimes.com).
4. As NTT is a former monopoly, it owned all the major telecommunications infrastructure. Given the cost of construction in Japan, new telecommunications companies piggyback on NTT's lines rather than constructing their own lines into households and office buildings. For those piggybacking rights NTT has been charging its would-be competitors an enormous amount ('four times the going rate in the US to interconnect to its lines' according to Catherine Pawasarat writing in the Oct. 2000 edition of *J@pan*

Inc. pp. 33–4.) which is then passed on to the consumer. As NTT also charges consumers Y10 a minute for 3 minute local calls, it is easy to see why Japanese Internet users have faced the highest costs worldwide for going on line. This is changed, however, as NTT cut the rate charged to its competitor telcoms by 22.5 per cent in 2002 so customers are likely to soon see lower Internet fees.

5. One very useful way of tracking current and future developments in Japanese m-commerce is through the work of the Mobile EC Subcommittee of the Electronic Promotion Council of Japan, established in April 2000 to encourage the expansion of e-commerce and to bring government and business together to work on problems and opportunities. ECOM is a consolidation of three other organizations; ECOM, Centre for the Informatization of Industry, and the Japan EC/CALS Organization. Details can be found in the *ECOM Journal*, at www.ecom.or.jp

6. This portable system works on a large number of small, lower power transmitters, located on buildings through major cities.

7. For an earlier overview of the development of mobile internet in Japan, see J.L. Funk, *Mobile Internet: How Japan Dialed Up and the West Disconnected* (ISI Publications, 2001).

8. Various terms are used to describe the mobile phone-based internet including mobile internet, wireless web, mobile e-services and mobile on-line services.

9. For a very general description of the evolution and operations of DoCoMo, see, John Beck and Mitchell Wade, *Docomo: Japan's Wireless Tsunami: How One Mobile Telecom Created a New Market and Became a Global Force* (Amacom, 2002). For a detailed assessment of DoCoMo's business plan, see J. Ratliff, 'NTT DoCoMo and its I-mode success: origins and implications', *California Management Review*, 2002, 44(3), p. 55.

10. DoCoMo was not the only company with the technology. Three other firms, J-phone, DDI Pocket Inc. and Astel, were in the market at the same time; all remain players in the mobile Internet field.

11. Shapiro, Elizabeth, 'Mari Matsunaga. reinventing the wireless web: the story of DoCoMo's i-mode', (http://www.japansociety.org/corpnotes/111400.htm).

12. Natsuno, Takeshi,. *i-mode Strategy* (Nikkei Business Publications Inc., 2000).

13. Matsunaga, Mari, *The Birth of i-mode* (Singapore: Chuang Yi Publishing, 2001).

14. Rohwer, Jim, 'No.1 Mari Matsunaga, 46: designer i-mode editor-in-chief e-woman', *Fortune Magazine*, Oct. 2000 issue (http:www.business2.com/articles/mag/0,1640,8612,00.html); Shapiro, Elizabeth. 'Mari Matsunaga. reinventing the wireless web: the story of DoCoMo's i-mode' (http://www.japansociety.org/corpnotes/111400.htm).

15. Rohwer, Jim, 'No.1 Mari Matsunaga, 46: designer i-mode editor-in-chief e-woman', *Fortune Magazine*, Oct. 2000 issue (http:www.business2.com/articles/mag/0,1640,8612,00.html).

16. Global Ethnographic Study of Wireless Use reveals key lessons for companies making wireless products and services: http://www.context/newsroom_article.cfm?ID=22

17. '2001 iForce Heroes: Kei-ichi Enoki, Managing Director, i-mode, NTT DoCoMo', iForce Initiative Feature Story (http:www.sun.com/2001-0829/feature/profiles/enoki.html).

18. As of Feb. 2001, along with the 300 yen monthly fee, DoCoMo charged 0.3 yen per packet of information. KDDI charged a monthly fee of 200 yen and 0.27 yen per package. J-Phone does not charge a monthly fee but charges two yen per information request.

19. Rose, Frank, 'Pocket Monster', *Wired*, Sept. 2001, p. 128.

20. 'Mobile phone net users to rival computer net users', *The Nikkei Industrial Daily*, Wednesday edn, 31 Jan. 2001(online edition).

21. Yutaka Mizukoshi, Kimihide Okino and Oliver Tardy, 'Lessons from Japan', *Telephony*, 15 Jan. 2001, p. 95.

22. For a detailed analysis of the potential of the keitai in this area, see H. Tsuji *et al.*, 'Spatial information sharing for mobile phones: digital cities II, computational and sociological approaches, 2002, Viol. 2362, pp. 331–42.

23. 'The internet, untethered: a survey of the mobile internet', *The Economist* Oct. 13, 2001, pp. 6–7.

24. 'On our Radar Screen', *J@pan Inc.*, June 2001, p. 62.

25 'The internet, untethered: a survey of the mobile internet', *The Economist* Oct. 13, 2001, p. 5.

26. Scuka, Daniel, 'Supplying new ideas: wireless lights up' *J@pan Inc.*, Nov. 2000, p. 44.

27. Nakayama, Shigeru, 'From PC to mobile Internet – overcoming the digital divide' Keynote Address, Internet and Society Conference, National University of Singapore, 14–15 Sept. 2001.

28. For an excellent summary of the commercial applications of the keitai, see Daniel Scuka, 'Unwired: Japan Has the Future in its Pocket', *J@panInc*, June 2000.

29. Tashiro, Chieko, 'Chieko's Diary' *J@pan Inc.* Oct. 2000, p. 8.

30. *J@pan Inc.*, May 2001, p. 7.

31. Frank Rose, 'Pocket monster: how DoCoMo's wireless internet service went from fad to phenom – and turned Japan into the first post-PC nation', *Wired*, Sept. 2001, p. 129.

32. Larimer, Tim, 'Internet a la i-mode', *Time Magazine*, 5 Mar. 2001, p. 54.

33. 'Japan's NPA panel drafts plan to ban Internet Dating sites', Xinhua News Agency, 26 Dec. 2002.

34. Larimer, Tim, 'Internet a la i-mode', Time magazine, 5 Mar. 2001, p. 54.

35. http://seattletimes.nwsource.com/news/nation-world/htm198/keit29_20000529.html

36. Scuka, Daniel, 'Supplying new ideas: wireless lights up', *J@panInc*, Nov. 2000.

37. Yutaka Mizukoshi, Kimihide Okino and Oliver Tardy, 'Lessons from Japan', *Telephony*, 15 Jan. 2001, p. 95.

38. 'E-commerce market seen to grow by 450% by 2005', Nikkei On-line: www.nni.nikkei.co.jp 7 Feb. 2001.

39. Neil Martin, 'Dialing for Yen', *Barron's*, Vol.78, no.42.

40. 'DoCoMo gets off to shaky 3G start', *The Nikkei Weekly*, 1 Oct. 2001, p. 8.

41. Hoffman, Andrea. 'The Other i-modes: Fifteen million happy non-DoCoMo users can't all be wrong', *J@pan Inc.* June 2001, p. 60.

42. Hoffman, Andrea. 'The other i-modes: fifteen million happy non-DoCoMo users can't all be wrong', *J@pan Inc.* June 2001, p. 60–1.
43. NTT DoCoMo: Business Success Stories: Http:www.nttdocomo.com/html/imode03_2.html
44. City Walk is a production of the Kobe Information and Multimedia Entertainment City, a government-sponsored initiative designed to encourage Kobe's economic involvement in the information revolution. See http://haikara.kimec.ne.jp
45. Abrahams, Paul. 'DoCoMo has an uphill battle in the US: population density, slow uptake hinder mobile use', *Financial Post (National Post)*, 22 Aug. 2000, p. C.10.
46. 'The internet, untethered: a survey of the mobile internet', *The Economist*, 13 Oct. 2001, p. 12.
47. For a somewhat less optimistic view of DoCoMo's overseas initiatives, see Daniel Scuka, DoCoMo, 'Easy Cell', *J@panInc.*, 17 Oct. 2000.
48. For details on the AOL and NTT DoCoMo agreement, see Nora Macaluso and Tim McDonald, 'AOL, NTT DoCoMo in wireless net pact', *E-Commerce Times*, 27 Sept. 2000.
49. Creed, Adam, 'Japanese carrier to send video to mobile phones': http:www.technews.com/news/00/158755.html
50. Yoshida, Junko, 'Java chip vendors set for cellular skirmish' at www.silicon strategies.com/story/OEE20010129S0063
51. Nakamoto, Michiyo, 'DoCoMo embarrassed by Java handset sales suspension', *The Financial Times*, 9 Feb. 2001; www.ft.com
52. Stephen Schwankert, 'Sun's shines on new DoCoMo services', http://asiaInternet.com/wireless/2001/02/0205-docomo.html
53. For an excellent international comparison of 3G implementations, see C. Banks 'The third generation of wireless communications: the intersection of policy, technology, and popular culture', *Law and Policy in International Business*, 2001, vol. 32, no. 3, 585–642.
54. 'NTT DoCoMo banks on 3G service', *The Nikkei Weekly*, 13 May 2002, p. 2; 'J-phone jumping into 3G market', *The Nikkei Weekly*, 9 Dec. 2002, p. 8.
55. 'Cell phone rivals play zero-sum game', *The Nikkei Weekly*, 13 Jan. 2003.
56. 'J-phone jumping into 3G market', *The Nikkei Weekly*, 9 Dec. 2002, p. 8.
57. 'Camera-equipped handsets snapped up', *The Nikkei Weekly*, 20 Jan. 2003.
58. 'Simpler handsets enjoy wide appeal', *The Nikkei Weekly*, 2 Dec. 2002, p. 8.
59. 'DoCoMo to chip in half to develop third-generation cell phones', *The Nikkei Weekly*, 20 Jan. 2003, p. 8.
60. Richard Meyer, Kei(tai)retsu, J@panInc. magazine, January 2002; http://www.japaninc.net/article.php?articleID=665
61. Mobile ED Subcommittee Report, *ECOM Journal No. 1* (www.ecom.or.jp)
62. 'Versatile cell phone'\, Japanese Companies at Home, *The Nikkei Weekly*, 30 Dec. 2002 and 6 Jan. 2003, p. 8.
63. 'World of cell phones widens', *The Nikkei Weekly*, 26 Aug. 2002, p. 10.
64. Schmetzer, Uli, *Chicago Tribune*, 'Computer? Who needs it? Japanese teens have I-mail': http://seattletimes.nwsource.com/news/nation-world/html98/keit29_20000529.html
65. 'Home appliances to get online links to cell phones', *The Nikkei Weekly*, 25 Nov., 2002, p. 10.

66. 'Feeding your pet through your cell phone', *J@pan Inc.*, Jan. 2003, p. 5.
67. NTT DoCoMo President's speech to the University of Berkley alumni association, June 2001, Tokyo; 'The internet, untethered: a survey of the mobile internet', *The Economist* 13 Oct. 2001, p. 16.
68. 'Information finding embodiment', *The Nikkei Weekly*, 4 Nov. 2002, p. 10.
69. On the transferability of the ketai revolution to the USA, see 'Wireless Internet – coming to Americad', *Electronic Business*, 2001, vol. 27, no. 10, p. 69.
70. 'The entertaining way to m-commerce: Japan's approach to the mobile Internet', *Electronic Markets*, 2002, vol. 12, no. 1, p. 6.
71. According to a November 2000 report, the Electronic Commerce Promotion Council of Japan suggested that the impact on GDP of e-commerce could be up to 13% over a 5–10 year period. They also noted that business to consumer e-commerce would represent only 2% of the market by 2004. M. Mahoney, 'Japan passes e-commerce catch-up bill', *E-Commerce Times*, 13 Nov. 2000 (Www.ECommerceTimes.com).
72. Larimer, Tim, 'Internet a la mode', *Time magazine*, 5 Mar. 2001, p. 54.
73. 'State want fast L-mode introduction', *Japan Times*, 3 Feb. 2001; 'Ministry postpones decision on NTT Internet service', Japan Times, 1 Feb. 2001.
74. 'L-mode: frequently asked questions', *Eurotechnology – Japan*, http://www.eurotechnology.com/Lmode/
75. 'Stimulus policy should continue', *Japan Times*, 21 July 2000.
76. Other countries are, of course, moving into the world of mobile Internet. In North America, one of the best systems is Research in Motion's Blackberry, which provides wireless Internet and email services. The quality, speed and practicality of the Blackberry pales in comparison to the Japanese services, and the m-commerce applications in North America are minimal.
77. Daniel Scuka, 'Supplying ideas: wireless lights up', *J@panInc*, Nov. 2000.
78. For a useful commentary on the difficult of transferring the Japanese m-commerce model to North America, see Mick Brady, 'Out on a limb with m-commerce', *E-Commerce Times*, 22 Sept. 2000.
79. For one analyst's view of the immediate future, see Renfield Kuroda, 'Wireless Predictions for 2001', *J@pan Inc.*, 15 Dec. 2000.
80. Richard Meyer, 'Kei(tai)retsu', *J@panInc.*, Jan. 2002: http://www.japaninc.net/article.php?articleID=665

4. Japanese E-Commerce

1. For a detailed and insightful study of the international development of e-commerce, see Lakshmi Iyer, Larry Taube and Julia Raquet, 'Global e-commerce: rational, digital divide and strategies to bridge the divide', *Journal of Global Information Technology Management*, Vol. 5, Issue 1 (2002).
2. Hiroski Araki, 'Expectations to the new ECOM', *ECOM Journal* 1(Http://www.ecom.jp/).
3. 'A new Japan', *Business Week – European Edition,* 25 Oct. 1999, p. 54.
4. 'Tokyo's valley of the netrepreneurs', *Business Week*, 6 Sep. 1999.
5. 'Japan: the Internet', *Business Week – European Edition*, 7 Feb. 2000, p. 1, 25.
6. 'Hiroshi Mikitana', *Business Week*, 14 Jan. 2002.

7. M. Lynskey and S, Yonekura, 'Softbank: an Internet Keiretsu and its leverage of information asymmetrics', *European Management Journal,* 2001, vo. 19, no. 1, pp. 1–15.
8. 'How son captured Japan's internet economy', *Fortune,* 16 Aug. 1999..
9. 'The last true believer', *Business Week*, 22 Jan. 2001.
10. 'Japan's web spinners: netrepreneurs are out to win on a global scale – and bring profound change', *Business Week,* 13 Mar. 2000.
11. 'The last true believer', *Business Week*, 22 Jan. 2001.
12. 'Japan's net generation', *Business Week*, 19 Mar. 2001.
13. Ibid.
14. For further details on Crayfish, see 'Deal mechanic', *Asiamoney*, May 2000, p. 11.
15. 'Japan.com fever', *Barron's*, 27 Mar. 2000.
16. Ibid. See also 'Promoting Japanese venture businesses', *Journal of Japanese Trade and Industry*, 2000, vol. 19, no. 4, p. 9.
17. 'Web shopping in Shinjuku', *Business Week*, 11 Dec. 2000.
18. E-Bay drops on mixed analysts view, *E-Commerce News*, 28 Feb. 2002.
19. 'Japan goes Web crazy', *Fortune,* 7 Feb. 2000.
20. 'Japan goes Web crazy', *Fortune,* 7 Feb. 2000.
21. 'Japan's Web spinners', *Business Week*, 13 Mar. 2000.
22. 'Japan: Fujitsu', *Business Week – European Edition*, 1999, 16 Aug. p. 27.
23. 'A Behomoth of the Net', *Business Week*, 23 Aug. 1999.
24. 'Amazon makes Japanese debut', *E-Commerce Times*, 1 Nov. 2000.
25. *Japan Internet report*, no. 66 (autumn 2002).
26. Building2Information Group, *U.S. online retailers in Japan* (http:www. builing2. com).
27. 'Wired for profit', *AdWeek*, 26 May 1997.
28. 'Retail Therapy', *Far Eastern Economic Review*, 28 Feb. 2002.
29. 'Bit valley grows up', *JapanInc.*, May 2001.
30. 'The last true believer', *Business Week*, 22 Jan. 2001.
31. 'After the crash', *Forbes,* 25 Dec. 2000.
32. 'View from the top', *Financial Times*, 6 Sept. 2000.
33. 'Show time at Softbank', *Business Week*, 10 Dec. 2001.
34. 'Softbank doubles down', *The Industry Standard*, 4 Sept. 2000.
35. 'Softbank to invest in Dotcom companies in Southeast Asia', Xinhua News Agency, 11 Mar. 2001.
36. 'Softbank', *Business Asia*, Dec. 2001, vol. 9, no. 11; *Mew Media Age*, 27 Sept. 2001.
37. 'After the party', *The Economist*, 14 Oct. 2000.
38. Neil Martin and Ken Belson, 'Japan.com fever', *Barron's*, vol. 80, no. 13, 27 Mar. 2000, pp. 26–9
39. Kitty McKinsey, 'Asians miss the e-biz mark', *Far Eastern Economic Review*, vol. 164, issue 10, 15 Mar. 2001.
40. Ken Belson, 'Net shopping: why Japan won't take the plunge', *Business Week*, vol. 3692, 31 July 2000.
41. *The Global Internet 100 – Survey 1988: Special Report*, 1998: http:// elabin-sead.edu/pdf/Global Internet100Survey2000.pdf
42. For an interesting comparison between Japanese convenience-store based e-commerce and American e-commerce, see Yuko Aoyama, 'Structural foun-

dations for e-commerce adoption: a comparative organization of retail trade between Japan and the United States', *Urban Geography*, 2001, vol. 44, no. 2, pp. 130–53.

43. 'Convenience Stores enjoy Net boost', *Nikkei Weekly*, 8 July 2002, p. 9.

44. Eddie Cheung, 'The land of rising ecommerce', Mar. 2001: emarketer.com

45. Eddie Cheung, 'The state of Asian ecommerce', Nov. 2000: emarketer.com.

46. The Economist Intelligence Unit/Pyramid Research Unit, *E-Readiness Rankings*, 8 May 2001.

47. There is a prevailing sense that the Japanese commercial culture is not well-suited to the pace and creativity of the 'new economy', an idea that has not totally disappeared with the collapse of the Internet bubble in North America. There is considerable evidence that certain kinds of inter-net development does well in a Japanese business environment. See S. Casper and H. Glimstedt, 'Economic organization, innovation systems, and the Internet', *Oxford Review of Economic Policy*, 2001, vol. 17, Issue 2, 265–81.

48. In Canada in 2001, a major debate broke out within the governing Liberal Party over a plan to implement a broadband Internet system for the entire country, at a cost running into the billions of dollars. When the leading advocate of the initiate, Industry Minister Brian Tobin, lost the battle, he resigned from the federal cabinet and left politics.

49. 'Japanese jump on interactive Web TV', *E-Commerce Times*, 3 July 2000.

50. 'A digital counterattack', *Forbes*, 18 Oct. 1999.

51. 'Reinvented; Japanese trading houses', *The Economist*, 5 May 2001.

52. 'Japan's Internet auction sites', *Far Eastern Economic Review*. 2000, vol. 163. No. 45, p. 51.

53. Ken Belson *et al.*, 'Asia's Internet deficit', *Business Week*, issue 3704, 23 Oct. 2000.

54. 'Localization: relating to customers around the world', 13 Sept. 2001: & CRMDaily.com

55. 'China B2B Web Site operator to start services in Japanese', *Nikkei Weekly*, 26 Aug. 2002, p. 20.

56. This discussion of Seven-Eleven is based on 'Over the counter e-commerce', *The Economist*, 26 May 2001.

57. 'Japanese banks embrace the e-future', *Institutional Investor*, 2000, vol. 25, no. 7, p. 48, 'Banks in Japan are finally embracing the Internet age', *Banker*, 2000, vol. 150, no. 890, p. 50'.

58. Uvan Schneider, 'Sakura Bank opens an Internet-only bank in Japan and plans to back another', *Bank Systems and Technology*, vol. 38, no. 1, Jan. 2001.

59. Yoshiaki Kiyota, 'Japan's publishing distribution in the Internet age', *Publishing Research Quarterly*, 2001, vol. 17, no. 2, p. 43.

60. 'E-Book market rides broadband to growth', *Nikkei Weekly*, 8 July 2002, p. 16.

61 'In a bind', *The Economist*, 30 Sept. 1999.

62. 'In Search of Doctor-San', *Business Week*, 11 Dec. 2000.

63. 'Peoplesoft defies sagging economy with strong Q3', CRM Daily.com, 19 Oct. 2001
64. 'The Internet, cars and DGPS – bringing mobile sensors and global correction services on-line, *GPS World*, 1 May 2000.
65. Ken Belson, 'Web shopping in Shinjuku', *Business Week*, issue 3711, 11 Dec. 2000.
66. Hiromi Ohki, 'IT contributes to growth in global trade', *Focus Japan*, vol. 27, issue 10, Dec. 2000, 8–9.

5 The Digital Face of Japan: National Dimensions of the Internet Revolution

1. 'Japan's Internet connection', *Focus Japan*, vol. 23, no. 1,2, Jan./Feb. 1996, pp. 1–2.
2. Tanya Clark, 'Asia wobbles onto the Web', Industry Week, vol. 246, no. 8, 21 Apr. 1997, pp. 80–82.
3. L.C.M. Nishioka, 'Special report: Japanese web sites target young working women: ASCII website (www.ascii.co.jp/english/news/archiv/98/01/21#1)'. Japanese women favor local speciality foods in online shopping, *Asia Biz Tech*, 24 Sept. 1998.
4. Glenn Hoetker, 'Konnichi wa, Nihon (Hello, Japan!)', *Database*, June 1994.
5. 'South Korea dominates Asia in Internet use', *Nielsen/NetRatings*, 14 March 2001.
6. *Digital Media Access Rate by Time of Day* (report by Jupiter Media Matrix), available at www.intage.co.jp/express/01_01/market/index1.html.
7. Digital behaviour of Japanese consumers, Oct. 2001: Http://www.intage.co.jp/express/01_01/market/index1.html.
8. 'Digital media behaviour of Japan consumers', Oct. 2001. Http://www.intage.co.jp/express/01_10/market/index1.html
9. 'Korean IT firms gain foothold in Japan', *Nikkei Weekly*, 1 Jan. 2001.
10. OECD, *Communications Outlook, 2001* (OECD, 2002).
11. Other countries, like Myanmar, Middle Eastern nations, and a few others, attempted to regulate the Internet. For some, including Myanmar, the concern was primarily political; the government did not want its citizens to gain unrestricted access to the outside world. For others – Singapore, Saudia Arabia, the United Arab Emirates – governments wished to ban material (pornography, gambling, and retail cites) that was not otherwise available in their country. The debate about national control of the Internet continues to rage, including efforts in Europe to regulate the sale of Nazi paraphernalia through e-commerce web-sites.
12. Ken Sakamura, 'Society needs a vision: progress alone won't be enough', *Japan Times*, 1 Jan. 2001.
13. Kathy Wilhelm, 'Tongue-tied on the Net', *Far Eastern Economic Review*, vol. 164, no. 7, 22 Feb. 2001.
14. Research on this subject remains embryonic. For an illustration of the importance of culture to Internet use in Japan and the USA, see Carrie La

Ferle *et al.*, 'Internet diffusion in Japan: cultural considerations', *Journal of Advertising Research*, 2002, vol. 42, Issue 2, pp. 65–79.

15. 'Osaka start-Up weaves Web of services', *Nikkei Weekly*, 7 May 2001.
16. For a discussion of the role of culture in the development of e-commerce, albeit with little reference to Japan, see L. Iyer *et al.*, 'Global e-commerce: rationale, digital divide, and strategies to bridge the divide', *Journal of Global Information Technology Management*, 2002, vol. 5, no. 1, p. 43. As this relates to Internet marketing, see R. Tian and C. Emery, 'Cross-cultural issues in Internet marketing', *Journal of American Academy of Business*, 2002, vol. 1, no. 2, pp. 217–24.
17. C. La Ferle *et al.*, 'Internet diffusion in Japan: cultural considerations', *Journal of Advertising Research*, 2002, vol. 4, no. 2, pp. 65–79.
18. One of the very best English-language sources on Japan available on the web is the *Kodansha Encyclopaedia of Japan Online*, available by subscription at www.ency-japan.net

6 Reflections on a Networked Japan: Japan and the Future of the Digital Revolution

1. While there is abundant commentary on e-commerce and m-commerce in Japan, and considerable discussion about technological aspects of the Japanese Internet, detailed analysis of the Internet cultural and social environment remains relatively limited. Some of the more insightful sources include William Underwood, 'Embracing cyberspace: the evolution of Japan's Internet culture', *World and I*, 2002, vol. 17, no. 6, p. 265; M.J. McClelland, 'Virtual ethnography' using the Internet to study gay culture in Japan', *Sexualities*, 2002, voo. 5, no. 4, p. 87.
2. Tioru Nishigaki, 'A Global Electronic Community: from the Fifth Generation Computer to the Internet', *Social Science Japan Journal*, 1998, vol. 1, no. 2, pp. 217–32.
3. R. McLaughlin, 'The Internet and Japanese education: The effect of education policies and government initiatives', *ASLIB Proceedings*, 1999, 51(7), pp. 224-32.
4 . The Internet, cars and DGPS', *GPS World*, 200, vol. 11, no. 5, p. 38. See also 'Digital wheels', *Business Week–European Edition*, 2000, April 10, p. 48.
5. 'Cars for the m generation', *J@pan Inc.*, Jan. 2003, pp. 8–9.
6. Shigeru Nakayama, 'From PC to mobile Internet – overcoming the digital divide in Japan', *Asian Journal of Social Science*, 2002, vol. 30, n. 4, pp. 239–47. This was not true before the advent of the keitia, when social class and occupation played, in Japan as elsewhere, a crucial role in determining access to technology. See T. Y. Morris-Suzuki and P. Rimmer, 'Cyberstructure and social forces: The Japanese Experience', *Sociologica*, 1997, vol. 12, no. 35, pp. 63–87.
7. 'Cyber-savy', *Nikkei Weekly*, 15 July 2002.
8. P. Rimmer and T. Morris-Suzuki, 'The Japanese Internet: visionaries and virtual democracy', *Environment and Planning*, 1999, vol. 31, no. 7, pp. 189–206.

9. I. Tweddell, 'The use of the Internet by Japanese new religious movements', MA thesis, University of Toronto, 2000.
10. For an excellent series of articles on the differential evolution of the Internet in Asia, see S. Rao and B. Klopfenstein (eds), *Cyberpath to Development in Asia: Issues and Challenges* (Westport: Praeger, 2002).
11. Manuel Castells, *End of Millennium* (New York: Blackwell, 2000), pp. 248–50.
12. The following is summarized from Manual Castells, *End of Millennium* (New York: Blackwell, 2000), pp. 251–5.

Select Bibliography

Abbate, J., 'Government, business and the making of the Internet', *Business History Review*, 2001, vol. 75, no. 1, pp. 147–76.

Anchordoguy, M., *Computers Inc.: Japan's Challenge to IBM*, Cambridge: Harvard University Press, 1989.

——, 'Japan at a technological crossroads: does change support convergence theory', *Journal of Japanese Studies*, 1997, vol. 23, no. 2.

Anchordoguy, M., 'Nippon Telegraph and Telephone Company (NTT) and the building of a telecommunications industry in Japan', *Business History Review*, 2001, vol. 75, no. 3, pp. 507–41.

Aoyama, Y., 'Structural foundations for e-commerce adoption: a comparative organization of retail trade between Japan and the United States', *Urban Geography*, 2001, vol. 44, no. 2.

Baber, Z., 'The Internet and social change: key themes and issues', *Asian Journal of Social Science*, 2002, vol. 30, no. 2, pp. 195–8.

Banks, C., 'The third generation of wireless communications: the intersection of policy, technology, and popular culture', *Law and Policy in International Business*, 2001, vol. 32, no. 3.

Beck, J. and Mitchell Wade, *Docomo: Japan's Wireless Tsunami: How One Mobile Telecom Created a New Market and Became a Global Force* (Amacom, 2002).

Brookings Task Force on the Internet, *The Economic Payoff from the Internet Revolution*, Washington: Brookings, 2001.

Casper, S. and H. Glimstedt, 'Economic organization, innovation systems, and the Internet', *Oxford Review of Economic Policy*, 2001, vol. 17, Issue 2.

Castells, M., *End of Millennium*, New York: Blackwell, 2000.

——, *The Rise of the Network Society*, New York: Blackwell, 1996.

Chang-tseh Hsieh, 'Japan's quest for global leadership in IT: the impact on US computer companies', *Industrial Management*, vol. 94, no. 2, 1994, pp. 23–28.

Chandler, A., *Inventing the Electronic Century: The Epic Story of the Consumer Electronics and Computer Industries*, New York: New Press, 2001.

Coates, K., 'Back in the race: Japan and the internet', *Japan after the Economic Miracle: in Search of New Directions*, P. Bowles and L. Woods (eds), Kluwer Academic Press, 2000, pp. 71–84.

Cusamano, M., *The Japanese Automobile Industry: Technology and Management at Nissan and Toyota*, Cambridge: Harvard University Press, 1985.

——, *Japan's Software Factories: A Challenge to U.S. Management*, New York: Oxford University Press, 1991.

——, *Thinking Beyond Lean: How Multi-Project Management Is Transforming Produce Development at Toyota and Other Companies*, New York: Free Press, 1998.

Dertousos, M., *What Will Be: How the New World of Information Will Change Our Lives*, New York: HarperBusiness, 1998.

Eades, J. *et al.*, *Globalization and Social Change in Contemporary Japan*, Melbourne: Trans Pacific, 2000

Fingleton, E., *Blindside: Why Japan Is on Track to Overtake the U.S. by the Year 2000*, Boston: Houghton Mifflin, 1995

——, *In Praise of Hard Industries*, Boston: Houghton Mifflin, 1999

Forester, T., *Silicon Samurai: How Japan Conquered the World's IT Industry*, Cambridge: Blackwell, 1993.

Fransman, M., *The Market and Beyond: Cooperation and Competition in Information Technology Development in the Japanese System*, Cambridge: Cambridge University Press, 1990.

Funk, J.L., *Mobile Internet: How Japan Dialed Up and the West Disconnected*, Hamilton, Bermuda: ISI Publications, 2001

Gates, B., *Business @ The Speed of Thought*, New York: Warner, 2000.

Gregory, G., *Japanese Electronics Technology: Enterprise and Innovation*, New York: John Wiley, 1986.

Hayes, D. and Martyn Warwick, 'Building bridges to the future', *Communications International*, vol. 24, no. 4, Apr. 1997, pp. 8–16.

Howell, T., 'The Internet as cure for Japan's ills', *World Trade*, 2001, vol. 14, no. 7.

Ieo, I., 'Being digital in Japan: the current Internet fever and multimedia in Japan', *Journal of Japanese Trade and Industry*, Nov. 1995, vol. 14, no. 6.

——, 'The state of digitization in modern Japan: the road to an advanced information network society', *Journal of Japanese Trade and Industry*, 1 Jan. 1997, vol. 16, no. 1.

Imai, K., 'The Japanese pattern of innovation and its evolution', in Nathan Rosenberg, R. Landau, and David Mowery (eds), *Technology and the Wealth of Nations*, Stanford: Stanford University Press, 1992, pp. 225–46.

Iyer, L., Larry Taube and Julia Raquet, 'Global e-commerce: rational, digital divide and strategies to bridge the divide', *Journal of Global Information Technology Management'*, 2002 vol. 5, issue 1.

Johnson, C., *MITI and the Japanese Miracle: The Growth of Industrial Policy*, Stanford: Stanford University Press, 1982.

Johnstone, B., *We Were Burning: Japanese Entrepreneurs and the Forging of the Electronic Age*, New York: Basic Books, 1999.

Kiyota, Y., 'Japan's publishing distribution in the Internet age', *Publishing Research Quarterly*, 2001, vol. 17, no. 2.

Kobayaski, K., *The Rise of NEC: How the World's Greatest C&C Company is Managed*, New York: Blackwell, 1991.

Kumagai, F., 'Possibilities for using the Internet in Japanese education in the information age society', *International Journal of Japanese Sociology*, 2001, no. 10.

La Ferle, C. *et al.*, 'Internet diffusion in Japan: cultural considerations', *Journal of Advertising Research*, 2002, vol. 42, issue 2.

Lambert, Bruce Henry, 'Building innovative communities: lessons from Japan's science city projects', European Institute of Japanese Studies, *Working Paper 107*, Nov. 2000.

Lynskey, M. and S. Yonekura, 'Softbank: an Internet Keiretsu and its leverage of information asymmetric', *European Management Journal*, 2001, vol. 19, no. 1.

McClelland, M., 'Virtual ethnography: using the Internet to study gay culture in Japan', *Sexualities*, 2002, vol. 5, no. 4.

McLaughlin, R., 'The Internet and Japanese education: the effect of education policies and government initiatives', *ASLIB Proceedings*, 1999, 51(7), pp. 224–32.

Malecki, E., *Technology and economic development: the dynamics of local, regional and national change,* New York: Longman, 1991.

Matsumura, K. *et al.*, 'A network system of medical and welfare information service for the patients, their families, hospitals, local governments, and commercial companies in a medical service area', *Journal of Medical Systems*, 2002, 26(3),

Matsunaga, Mari, *The Birth of i-mode*, Singapore: Chuang Yi Publishing, 2001

Mizushima, et al., 'Japanese experience in telemedicine in oncology', *International Journal of Medical Informatics,* 2001, pp. 612–3.

Morita, A. with Edwin Reingold and Mitsuko Shimomura, *Made in Japan: Akio Morita and SONY ,* New York: Dutton, 1986.

Morris-Suzuki, Tessa, 'The "communications revolution" and the household: some thoughts from Japan', *Prometheus.*, 1988, 6. 2.

——, *The Technological Transformation of Japan*, Cambridge: Cambridge University Press, 1994.

——, and P. Rimmer, 'Cyberstructure and social forces: the Japanese experience', *Sociologica,* 1997, vol.12, no 35.

——, 'Cyberstructure, society and education: possibilities and Problems in the Japanese Context', National Institute of Multimedia Education, International Symposium, 1999.

Murai, Jun, *Intânetto sengen,* Tokyo: Kôdansha,1995

Nakayama, S., 'From PC to mobile Internet -– overcoming the digital divide in Japan', *Asian Journal of Social Science*, 2002, vol. 30, no. 4.

Natsuno, Takeshi, *I-mode Strategy ,* Nikkei Business Publications , 2000

Necroponte, N., *Being Digital ,* New York: Vintage, 1996.

Nishigaki, T., 'A global electronic community: from the fifth generation computer to the Internet', *Social Science Japan Journal,* 1998, vol. 1, no. 2

Noble, D., *Digital Diploma Mills: The Automation of Higher Education,* New York: Monthly Review Press, 2002.

Ozawa, T., 'The new economic nationalism and the "Japanese disease:" The conundrum of managed economic growth', *Journal of Economic Issues*, June 1996.

Perez P., 'The Internet phenomenon in Japan', *Communications et Strategies*, 1998, no. 32, p. =157.

Porter, M., Hirotaka Takeuchi and Mariko Sakakibara, *Can Japan Compete?.* London: Macmillan – now Palgrave Macmillan, 2000.

Posner, A., 'Japan', in B. Steil, David Victor and Richard Nelson, (eds), *Technological Innovation and Economic Performance,* Princeton: Princeton University Press, 2002, pp. 74–111.

Rao, S., and B. Klopfenstein, (eds), *Cyberpath to Development in Asia: Issues and Challenges*, Westport: Praeger, 2002

Ratliff, J., 'NTT DoCoMo and its i-mode success: origins and implications', *California Management Review*, 2002, 44(3).

Rimmer, P. and T. Morris-Suzuki, 'The Japanese Internet, visionaries and virtual democracy', *Environment and Planning*, 1999, Vol. 31, no. 7.

Steil, B., David Victor and Richard Nelson, (eds), *Technological Innovation & Economic Performance* , Princeton: University of Princeton Press, 2002.

Takahashi, T., 'The present and future of telemedicine in Japan', *International Journal of Medical Informatics,* 2001, vol. 61, no. 2.

Thompson, C.S., 'Enlisting on-line residents. Expanding the boundaries of e-government in a Japanese rural township', *Government Information Quarterly,* 2002, 19(2).

Tian, R., and C. Emery, 'Cross-cultural issues in Internet marketing', *Journal of American Academy of Business,* 2002, vol. 1, no. 2

Tilton, John, *International Diffusion pf Teechnology: Thee case of Semiconductors* , Washington: Brookings Institution, 1971

Tkach, L., 'Politics@Japan: Party Competition on the Internet in Japan', *Party Politics,* 2003 vol.9, no 1, pp. 105-23.

Tsuji, H. *et al,* 'Spatial information sharing for mobile phones', *Digital Cities II, Computational and Sociological Approaches,* 2002, Vol. 2362.

Tweddell, I., 'The use of the Internet by Japanese new religious movements', MA thesis, University of Toronto, 2000.

Vogel, E., *Japan as Number One: Revisited,* Singapore: Institute of Southeast Asian Studies, 1986.

Underwood, W., 'Embracing cyberspace: the evolution of Japan's Internet culture', *World and I,* 2002, vol. 17, no. 6.

Index